FUNDAMENTALS OF X-RAY PHYSICS

FUNDAMENTALS OF X-RAY PHYSICS

G.V. Pavlinsky

Viva Books

New Delhi | Mumbai | Chennai | Kolkata | Bangalore | Hyderabad | Kochi

First Indian Edition 2009

VIVA BOOKS PRIVATE LIMITED

- 4737/23 Ansari Road, Daryaganj, New Delhi 110 002
  E-mail: vivadelhi@vivagroupindia.net, Tel. 42242200
- 76, Service Industries, Shirvane, Sector 1, Nerul, Navi Mumbai 400 706
  E-mail: vivamumbai@vivagroupindia.net, Tel. 27721273, 27721274
- Jamals Fazal Chambers, 26 Greams Road, Chennai 600 006
  E-mail: vivachennai@vivagroupindia.net, Tel. 28290304, 28294241
- B-103, Jindal Towers, 21/1A/3 Darga Road, Kolkata 700 017
  E-mail: vivakolkata@vivagroupindia.net, Tel. 22836381, 22816713
- 7, Sovereign Park Apartments, 56-58, K. R. Road, Basavanagudi, Bangalore 560 004
  E-mail: vivabangalore@vivagroupindia.net, Tel. 26607409, 26607410
- 101-102 Mughal Marc Apartments, 3-4-637 to 641, Narayanguda, Hyderabad 500 029
  E-mail: vivahyderabad@vivagroupindia.net, Tel. 27564481, 27564482
- Beevi Towers, First Floor, SRM Road, Kaloor, Kochi 682 018, Kerala
  E-mail: vivakochi@vivagroupindia.net, Tel: 0484-2403055, 2403056

www.vivagroupindia.com

Published by arrangement with
Cambridge International Science Publishing
7, Meadow Walk, Great Abington
Cambridge CB1 6AZ, UK

ISBN-10: 81-309-0947-2
ISBN-13: 978-81-309-0947-9

Published by Vinod Vasishtha for Viva Books Private Limited, 4737/23, Ansari Road, Daryaganj, New Delhi-110 002. Printed and bound by Anand Sons, Delhi-110 092.

Contents

PREFACE

Most of the phenomena relating to the physics of x-radiation were discovered and investigated in the first third of the 20th century. The relationships, determined in the 'golden age' of physics, represent the corner stone of all subsequent investigations in this area. They form the basis of considerable achievements in the investigations of synchrotron radiation, x-ray optics, Raman scattering, x-ray fluorescence, etc.

It is difficult to list all practical applications of x-radiation. The knowledge of the physical fundamentals of the formation of x-radiation and of its interaction with matter is essential for determining new and developing traditional directions of application in chemistry, biology, metallurgy, materials science and other branches.

I hope that the book will be useful to readers concerned with the investigations in the specific area of the physics of x-radiation or its application because knowledge of other directions enriches investigations and ensures the high quality of solving the given task. The material presented in the book will also be used to experts working in other areas of science, since the intersection of fruitful concepts is often the source of scientific advances. Finally, information on x-radiation and its properties is essential for graduates of physical specialisations, because there is a wide range of directions in which this knowledge will be used to various degrees.

The need for this book was dictated mainly by the fact that more than 50 years have passed since the publication of the last Russian monographs on this subject (Blokhin M.A., Physics of x-rays, 1957, Borovskii I.B., Physical fundamentals of x-ray spectrum investigations, 1956). Later editions of books concerned with individual sections and applications of x-radiation physics naturally contain only the information on these sections.

Physics of x-radiation is at present such a large and multifaceted topic that several volumes would be required for complete presentation, with the authors of the individual volumes being experts in various areas. This book describes only the fundamentals of the physics of x-radiation without any extensive mathematical substantiation of the results. This shortcoming is eliminated by references to fundamental studies in recent years, reflecting the most important achievements in each area. Naturally, the book unavoidably reflects the specialisation of the author. Therefore, the sections relating to x-ray fluorescence and the process of interaction of the electron beam with matter are investigated more extensively.

The book is based on a series of lectures by the author in the course of several years at the Irkutsk State University. As in the lectures, the material is presented in a sequence, using the results obtained in previous chapters in each new section.

I hope that the material presented in the book will be useful to a wide range of readers.

INTRODUCTION

Reports on the new, previously unknown type of radiation were published by Wilhelm Konrad Röntgen in December 1895 in the journal 'Sitzungsbericht Med. – Phys. Gesellschaft' [1]. The rays formed during the passage of electrical current ('cathode rays') through a rarefied gas in the area of incidence of the rays on a glass cylinder of a discharge tube. The new radiation, are referred to as x-radiation, had the following properties: propagation in a straight line, no effect of the magnetic field on radiation, glow of fluorescent screen under the effect of a this radiation screen, high penetrating capacity. Shadow photographs of a closed set of drawing instruments and of a hand wrist in the transmitted x-radiation were published in the article.

The interest in the Röntgen's discovery was enormous. More than a thousand studies in this area were published in the following 12 months. The importance of the discovery is indicated by the fact that in the period to 1927 six Nobel prizes were awarded for the investigations in the area of the new type of radiation. The first Nobel Prize in this area was awarded to W.K. Röntgen himself (1901), and subsequently to M. Laue (1914), W.G. and W.L. Bragg (1915), I. Barkla (1917), M. Siegbahn (1924) and A. Compton (1927).

Naturally, the primary task was the determination of the nature of x-radiation.

The task was complicated by the fact that at the time of discovery, the properties of cathode rays, causing this radiation, were not completely known. The charge and mass of the particles, representing the cathode rays and referred to as electrons, were determined in experiments carried out by Thomson only in 1897–1898.

In the investigations, an analogy was found between the properties of x-radiation and the electromagnetic radiation of the visible range of the spectrum. For example, in 1905, I. Barkla proved the polarisation of scattered x-radiation. However, a more significant confirmation of its electromagnetic nature was the diffraction of this radiation by a crystal lattice discovered by Max von Laue in 1912. Laue proposed to referred to x-radiation as röentgen radiation, although the previous term is still being used in a number of countries even now. The refraction of x-radiation, discovered by Stengstrem in 1917, confirmed the conclusion on the electromagnetic nature of this radiation.

The diffraction of x-radiation on crystal structures confirmed the fact that the wavelength of this radiation is comparable with the atomic spacing in the crystal lattice. In fact, according to the current views, the x-rays occupy a range from hundredths of a nanometer to tens of nanometres in the spectrum of electromagnetic radiation. This is consistent with the atomic spacing values.

The discovery of the electromagnetic nature of x-radiation represented the basis for further fundamental investigations in this area.

The most important stage of the investigations of x-radiation was the interference reflection of this radiation from atomic planes of ordered structures (crystals), discovered in 1913 by W.G. and W.L. Bragg. The relationship between the direction of the wave, reflected by the crystal, and the parameters of the crystal lattice was discovered simultaneously by Yu.V. Wulf independently of W.G. and W.L. Bragg. Since the x-radiation could be decomposed into a spectrum, it was possible to separate linear and continuous components of the composition of the spectrum. It is important to say that the presence of the linear component was discovered earlier by I. Barkla (in 1908) when comparing the experimental data on the absorption of radiation of x-ray tubes with different anodes. However, the extensive possibilities of spectral separation and examination of the lines of x-radiation became available only after derivation of the Wulf–Bragg equation. This tool enabled Moseley to determine, already in 1913, the relationship between the atomic number of the element and the wavelength of linear x-radiation typical of this element (Moseley law). Moseley also showed that there were at that time six undiscovered elements with the atomic numbers 43, 1, 72, 75, 85 and 87 between hydrogen and uranium. Using the Moseley law, G. Hevesy and D. Koster discovered in 1923 the 72nd element of the periodic system of elements (hafnium), and in 1926 V. and I. Noddak discovered the 75th element of this system (rhenium). The linear spectrum was referred to as characteristic because it characterises the element emitting this spectrum.

The mechanism of formation of characteristic x-radiation was proposed in 1916 by W. Kossel who used the theory of the structure of the atom proposed by Neil Bohr. The improvement of Bohr's theory was accompanied by progress of understanding on the spectral composition of characteristic radiation and the relative intensity of the lines of this spectrum. Working in this direction, A. Sommerfeld, H. Wentzel, D. Koster and others carried out a general systematisation of the characteristic x-ray spectrum.

A considerable amount of work in the measurement of the wavelength of characteristic x-ray radiation of different chemical elements was carried out in the period from 1913 to 1923 by M. Siegbahn on the vacuum x-ray spectrometer which he constructed and which was unique at that time.

The first reports on the spectral distribution of continuous x-radiation were obtained by Moseley and Darwin and in 1913 by the decomposition of the radiation of an x-ray tube into a spectrum. Subsequent investigations of the dependence of the spectral distributions on the excitation conditions were conducted by S.T. Ulrey, N. Behnken, A. March and a number of other investigators. In 1922, H. Kullenkampff used the available and his own empirical data to derive an analytical equation for the spectral intensity of continuous x-radiation and its dependence on the voltage on the tube and the atomic number of the anode material. A year later, H. Kramers developed the theory of formation of this radiation, examining the process of deceleration of the electrons in the field of the atomic nucleus. The theoretical results published by Kramers were in good agreement with the Kullenkampff empirical equation. Continuous x-radiation is often referred to as bremsstrahlung in accordance with its formation mechanism.

The quantum nature of the electromagnetic radiation was efficiently validated by the theory of noncoherent scattering of x-radiation, proposed by Compton in 1923.

X-ray fluorescence was discovered in 1908 by Barkla and Sandler. They showed that the hardness of fluorescent radiation, emitted by the irradiated objects, is the characteristic of these objects. Later (1914) de Broglie published the results of investigations of x-ray fluorescence, excited by the radiation of an x-ray tube. In 1932, Hevesy and Koster used fluorescent x-ray radiation for increasing the contrast of the analytical signal in the detection of characteristic lines of hafnium. At present, the x-ray fluorescence method is one of the main methods in analytical chemistry.

Thus, at the start of the 30s of the previous century, the stage of development of the theoretical fundamentals of the formation of x-radiation and of investigations of its fundamental properties was practically completed.

The period from the 30s to 60s of the 20th century was the period of extensive introduction of the x-ray methods of investigations of matter and materials in different areas of science and production. This was the time of construction of sensitive x-ray equipment, improvement of the detectors of ionising radiation and detailed investigations of the x-ray spectrum, and the construction of new and increasing the power of existing sources of x-ray radiation, the development of the fundamentals of crystallography, x-ray spectroscopy, x-ray microscopy, x-ray structural and x-ray spectral methods of analysis, and the formation of other areas of application of this type of radiation.

Later, x-radiation was used on an increasing scale as an efficient tool for solving a large number of greatly varying problems in science and production. This period is characterised by the formation of instrumental complexes, using the latest achievements in instrument making, computing methods and mathematical apparatus for the processes of processing experimental information.

The theory of formation of x-radiation and of its interaction with matter was described extensively in the Russian and translated literature of the 40s to 60s (see, for example [2–7]), in monographs dealing with individual properties of this radiation (optics [8–10], scattering [11, 12], bremsstrahlung [13], x-ray spectrometry [14–17]). The theory was presented in the shortened form in books dealing with x-ray microprobe analysis [18–23], x-ray diffraction [24] and x-ray fluorescence analysis [25–28], and also with the application of x-ray fluorescence analysis in analysis of various materials [29–33].

When examining the fundamentals of the physics of x-radiation, it is necessary to take into account the more recent results, obtained in this area (thermal and synchrotron excitation of x-radiation, the construction of x-ray optical systems, development of the theoretical fundamentals of formation of x-ray fluorescence, etc,) avoiding if possible intermediate computations in the presentation of the final theoretical expressions. The book provides general information on this extremely important area of science and is intended for experts working with the applications of x-radiation.

Chapter 1

Characteristic x-ray radiation

Characteristics x-ray radiation is determined by the atom structure. Therefore, it is rational to start examination of the characteristic spectrum with examination of the atomic structure.

1.1. Development of concepts of the atom structure

The concept of the atom as the smallest indivisible particle of matter (atom is from the Greek word atomos, i.e. indivisible) was proposed a very long time ago. The first model which made it possible to estimate the atom dimensions was proposed by Thomson. According to this model, the atom is a sphere with a uniform charge in the volume. The charge of the sphere is equal to $+Ze$. In the centre of the sphere there are Z electrons. Therefore, on the whole, the atom remains neutral. Experiments with scattering of electrons on atoms with a comparably low energy indicating the constant dimensions of the atom and this corresponded to the considerations of the proposed model.

It was found that the Thomson model is unfounded when the effective scattering cross-section for the velocities of the scattered electrons, comparable with the velocity of light, started to decrease in proportion to the fourth degree of the electron velocity. This fact indicated that the sphere, limiting the size of the atom, is not filled with matter with a positive charge and with electrons in its centre, and, evidently, the negative charge is distributed in space in the form of a thin shell on the atom surface. This shell repulses the electrons with moderate velocity and is transparent to the electrons with high energy. The scattering of electrons with high energy shows that the positive charge of the atom is concentrated in the nucleus of the atom with the linear dimensions of the order of 10^{-11} cm.

The nuclear model of the atom was proposed by Rutherford [34]. According to this model, the atom consists of a positively charged nucleus in which almost the entire mass of the atom is concentrated, and of a shell formed by electrons moving around the nucleus on circular or elliptical orbits.

Experimental verification of the validity of the model was carried out by Rutherford by investigating the scattering of α-particles on atoms of heavy elements. It was assumed that the mass of the α-particles is small in comparison with the mass of the atomic nucleus and that the interaction between the α-particle (with the charge +2*e*) and the nucleus is determined by the Coulomb law. Since the charges of the nuclei and of the α-particles are identical, the trajectory of the α-particle is hyperbolic in which the angle ϑ of deviation from the initial direction is determined by the equation:

$$\operatorname{ctg}\left(\frac{\vartheta}{2}\right)=\frac{M_{\alpha}V^{2}}{2Ye}p,$$

where M_{α} is the mass of the α-particle, V is is velocity, Ye is the charge of the nucleus, p is the parameter which determines the proximity of the initial vector of the α-particle to the atom nucleus.

The experiments carried out by Rutherford confirmed the validity of the equation thus confirming the accuracy of the proposed model. Similar experiments enabled the charge of the nucleus Z to be determined. Measurements taken by Chadwick [35], showed that the charge of the nucleus is a multiple of the atomic number Z of the element in the periodic table of elements, i.e. $Y = Ze$, where e is the elementary charge

However, Rutherford's model was not in agreement with the considerations regarding the emission of energy by electrons travelling with acceleration in the field of the nucleus. Taking into account the laws of classical physics, a particle moving with acceleration is a source of electromagnetic radiation. Consequently, the total energy of the electron in the orbit should decrease and the rotation frequency of the electron around the nucleus should vary continuously. Therefore, the emission spectrum should be continuous and this did not correspond to the experimentally determined linear spectra of emission of the atoms. In addition to this, losing energy, the electron should, in the final analysis, fall on the nucleus and the atom should cease to exist. The incidence time of the electron is determined by a value of the order of 10^{-8} s.

A model of the atom which was in agreement with the experimental data was proposed by Bohr [36] in 1913. Examining the movement of the electrons around circular orbits, Bohr supplemented Rutherford's modelling considerations by the following two postulates:

1. Of all the circular orbits possible from the viewpoint of classical mechanics, a discrete set of orbits corresponding to the stationary state of the atom is formed. The electron on these stationary orbits does not emit electromagnetic waves, although its movement is accelerated.

2. The set of the stationary orbits corresponds to the discrete set of energies E_n of the electron in the field of the nucleus. In transition of the electron from one energy state to another, emission or absorption of a quantum takes place with the frequency:

$$\nu = \frac{E_{n1} - E_{n2}}{h}$$

where $n1$ and $n2$ are the numbers of the states, h is the Planck constant.

Bohr's concept of the existence of stationary energy states of the atomic systems found complete experimental justification. In particular, the calculated wavelength values of the characteristic lines were in almost complete agreement with the experimental values for the hydrogen atom and for a hydrogen-like, once-ionised helium atom.

Bohr's theory was developed further by Sommerfeld [3] who extended the conclusions of the theory to elliptical orbits, using canonic dependences of the classic mechanics and the quantum-mechanical model, proposed by Bohr. The quantisation of azimuthal moments, proposed by Sommerfeld, required introduction of the concept of the azimuthal quantum number, supplementing the main quantum number. The Bohr–Sommerfeld model also took into account the variation of the electron mass at high velocities of rotation around the nucleus.

The intrinsic momentum of the electron (its 'spin') was observed by Stern and Gerlach [37] in experiments with splitting of lines in a strong magnetic field.

Consequently, the following equation was derived for the energy of the level of the hydrogen-like atom:

$$E = -Rhc\frac{M_z}{M_z + m}\left(\frac{Z^2}{n^2} + \alpha^2\frac{Z^4}{n^4}\left(\frac{n}{j+1/2} - \frac{3}{4}\right)\right), \quad (1.1)$$

where $R = (2\pi^2\, me^4)/h^3c = 109737.303\text{cm}^{-1}$ is the Rydberg constant, $h = 6.6242\cdot10^{-34}$ J·s is the Planck constant, $m = 9.1066\cdot10^{-28}$ g is the mass of the electron, $c = 2.99776\cdot10^8$ m/s is the velocity of light, $\alpha = \frac{2\pi e^2}{hc} = 7.2981\cdot10^{-3}$ is the constant of the fine structure, n is the main quantum number, j is the internal quantum number which characterises the total angular and magnetic momentum of the electron taking into account its spin moment s, i.e. $j = l \pm s$, where l is the azimuthal quantum number.

Since the quantum spin number $s = \pm 1/2$, then $j = l \pm 1/2$. M_Z is the mass of the atomic nucleus.

Quantity $M_Z / (M_Z + m)$ reflects the fact that the electron, having the mass m, does not rotate around the nucleus but it rotates around the general centre of the masses $(M_Z + m)$.

The minus sign in equation (1.1) is determined by the selection of the zero level of energy which corresponds to the removal of the valency electron from the atom. Consequently, the energy of optical levels of the atom on which the transitions of the electrons are possible during excitation of the atom, should be regarded as negative.

The quantisation of the energy of the electron orbits in the Bohr-Sommerfeld model may be explained on the basis of the wave properties of the electron and by examining its orbital movement in the atom as a standing wave formed in a limited volume.

In 1926, E. Schrödinger [38] presented an equation for describing the condition of quantum objects using the wave function Ψ. The Schrödinger equation is the main dynamic equation of the non-relativistic quantum mechanics and expresses the corpuscular-wave dualism of microparticles. The so-called stationary Schrödinger equation has the form:

$$-\frac{h^2}{2m}\nabla^2\Psi + V(x,y,z)\Psi = E\Psi,$$

where m is the mass of the particle, moving under the effect of the potential $V(x, y, z)$, E is the total energy of the quantum system, ∇ (nabla) is the Hamiltonian operator.

The solutions of the Schrödinger stationary equation are available for discrete values of the energy E_n with the corresponding wave functions Ψ_n. When discussing the atomic systems, the exact solution of the Schrödinger equation is possible only for the one-electron atomic structures (hydrogen, helium ions He^+ and lithium ions Li^{++}). The determined energies of the levels of these structures are determined by the equation (1.1) without the second term in the square brackets.

A many-electron atom is a complicated system of interacting particles. Therefore, the wave function Ψ of a many-electron system, reflecting the condition of the system as a whole, depends on the coordinate and spin variables of all the elements of the system.

The sum of the operators of the individual fields and their interaction are determined by the Hamiltonian of the system. The Hamiltonian of a many-electron system contains mainly the kinetic energy of the electrons

$-\frac{h^2}{2m}\sum_i \nabla_i^2$, the energy of interaction of each electron with the potential, formed by the nucleus, and the energy of Coulomb interaction of all possible electron pairs. The last factor (the electron–electron interaction) does not make it possible to describe the wave function in the form of a product of the one-electron wave functions each of which would be a function of only the coordinates of one electron. To overcome this difficulty, Hartree [39] proposed a variation procedure in which the wave function is approximated by the product of the one-electron functions. This variation procedure results in the following Hartree equations written for the one-electron wave functions, minimising the energy of the system:

$$\left(-\frac{h^2}{2m}\nabla^2 + V(r) + \sum_{j\, j\neq i} e^2 \int \frac{\psi_j^*(r')\psi_j(r')dr'}{|r-r'|} \right) \psi_i(r) = \varepsilon_i \psi_i(r).$$

Every electron is subjected to the effect of forces, determined by the potential $V(r)$ which takes into account the Coulomb interaction of the electron with the nucleus and the average effect of all other electrons. This effect is given by some mean self-consistent field which depends on the selection of the one-electron wave functions, and these functions depend in turn on the strength of the mean field.

The values of ε_i in the Hartree equations are variation parameters of the intrinsic energies of the one-electron state.

The successive Hartree–Fock approximation increases the accuracy of the Hartree equation by taking into accoun the correlation of the orbits of the electrons, associated with the exchange interaction of the electrons.

In the case of a hydrogen-like atom, the Schrödinger equation gives the results which do not differ greatly from the results obtained in the Bohr-Sommerfeld quantum-mechanical approach. Since the model of the hydrogen-like atom has proved to be sufficient for the examination of the problems of formation of characteristic x-ray radiation, the relationship (1.1) is used as a basis for systematisation of the x-ray spectra. However, when examining some problems of the physics of x-ray radiation, it is necessary to use the considerations of the wave properties of the electron.

1.2. The equation for the energy of x-ray levels of the atom

The energy of some x-ray level i is the energy of the atom in which one electron is removed from this level.

According to the Pauli vacancy principle, the set of the possible values of the energies of the levels in the presence of Z electrons outside the closed shell of the atom is the same as in the presence of Z vacancies in an open shell. Therefore, the so-called hydrogen-like atom, with a single electron in an unfilled shell, gives the same set of the permissible values of the energies of the levels of the atom as the atom whose electron shell does not contain any electron. Consequently, the energies of the x-ray levels without the single electron in the inner shell are identical with the energies of the levels of the hydrogen-like atom.

Two special features should be noted in this case.

1. The zero x-ray level is the energy of the atom from which one valency electron is removed. Since the energy of the level increases with the decrease of the distance to the nucleus, the energy of the atom with the inner electron removed is positive and in the equation (1.1), written for the energy of the optical levels of the hydrogen-like atom, the sign for the x-ray levels should be changed to positive. In transition of the electron from some outer level f to the inner level i, the atom carries out transition from the state with a higher energy i to a state with lower energy f, i.e., the energy of the atom decreases.

2. The electron on the inner level is screened from the nucleus by all other electrons of the atom. Therefore, in the atomic number Z of the main (first) term of the equation (1.1) it is necessary to make a correction σ_1 for full screening (σ_1 is a full screening constant). Since the screening effect increases with an increase of the number of electrons in the atom and with an increase of the distance of the shell from the nucleus, the full screening constant σ_1 increases with the increase of Z, n and l.

The second term in equation (1.1) reflects the effect of the magnetic interaction between azimuthal and spin moments on the energy of the level. The close shells have zero magnetic moment. The open form of the shell, determined by the removal of one electron, results in the formation of its magnetic moment and, consequently, in the second term it is necessary to make the correction σ_2 for internal screening. Since the number of the electrons on the inner shells of the atom is constant (with the exception of the atoms with the smallest numbers Z), the internal screening constant σ_2 is independent of Z and increases with increasing n and l.

Taking this into account, using the relationship (1.1), the equation for the energy of the x-ray level is presented in the form:

$$E(n,l,j) = Rhc\frac{M_z}{M_z+m}\left(\frac{(Z-\sigma_1)^2}{n^2}+\frac{\alpha^2(Z-\sigma_2)^4}{n^4}\left(\frac{n}{j+\frac{1}{2}}-\frac{3}{4}\right)\right), \quad (1.2)$$

where σ_1 and σ_2 are the constants of the full and internal screening, respectively.

The main quantum number n = 1, 2, 3,... is determined by the shell to which the electron belongs. These shells in x-ray spectroscopy are denoted as follows:

n = 1 2 3 4 5 6

Shell: K L M N O P

The azimuthal quantum number l determines the orbital angular (and, consequently, magnetic) moment of the electron and characterises the state of the electron in the shell. These states are quantized and may vary in the range from 0 to $n-1$. The distribution inside the shell with respect to the state is denoted in x-ray spectroscopy by the following letters:

$l =$	0	1	2	3
State:	*s*	*p*	*d*	*f*

Taking these considerations into account, the following notations of the energy levels of the atom have been accepted in x-ray spectroscopy:

Type of state	1*s*	2*s*	2*p*	3*s*	3*p*	3*d*
Notation	K	L_1	$L_{2,3}$	M_1	$M_{2,3}$	$M_{4,5}$

The double indexes characterise the splitting of the appropriate level into two similar sublevels (section 1.3.5).

The energy levels of a free atom are shown in Fig. 1.1*a* (without maintaining the true relative distances between them). The outer shell of the atom, consisting of valency electrons, is denoted by letter V. The levels of excited states of the valency electrons and indicated by the thin lines and denoted by e. The region of continuous energy is situated above the valency levels.

The formation of characteristic x-ray radiation requires the removal of an electron from the inner shell into the continuous region of energies. This process of ionisation of the atom is represented by the transition 1, characterising the absorption (since the ionisation is associated with the absorption by the atom of the energy of the particle (or photon) that ionises the atom). The direction of the arrow characterises the transition of the atom from the non-excited state to the single ionisation state. The emission of an x-ray quantum takes place during the transition of the electron from

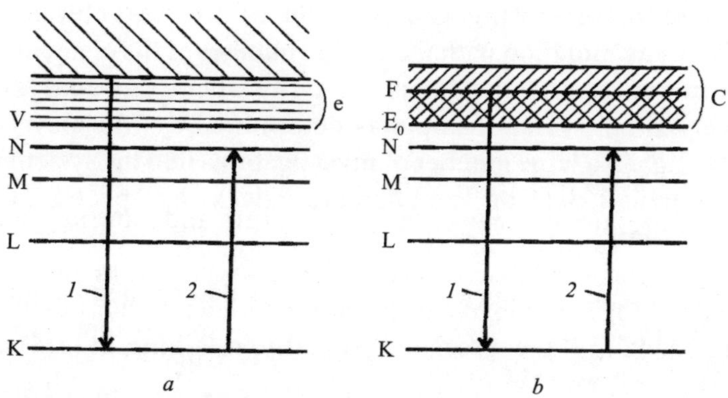

Fig. 1.1. Diagrams of the levels of the free atom *a* and electronic states of the metal *b*. 1) of the absorption process, 2) the emission process.

a remote shell in relation to the nucleus to a shell closer to the nucleus. This process is denoted by transition 2 characterising emission (the direction of the arrow characterises the transition of the atom from the K-ionisation to N-ionisation state).

If the atom is included in the composition of a solid, it should be taken into account that the valency electrons of the atoms form single energy bands of resolved and forbidden states, and the filling of these bands depends on the nature of electrical conductivity of matter. Therefore, the electron removed from the atom during ionisation of the latter may transfer into the unfilled zones of resolved states with a complicated structure. The simplest diagram of the levels of metal atom for a solid is shown in Fig. 1.1*b*. In this case, the outer electrons of the atoms efficiently screen the inner electron shells from the effect of adjacent atoms and, consequently, the energy levels in Fig. 1.1*a* and 1.1*b* coincide almost completely. For metals, the valency electrons merge into a conductivity band in which the resolved energies have a continuous spectrum. This band is indicated by letter C in Fig. 1.1*b*. Its lower level (the bottom of the conductivity band) is E_0. The electrons fill the conductivity band to the Fermi level, denoted by F. Above the Fermi level, the conductivity band is free and capable of accepting electrons formed during atom ionisation. This means that the energies of the particles (or photon), ionising the atom, should be sufficient for the transfer of the electron in the process of absorption into the free zone of the resolved states above the Fermi boundary. The process of emission for the metal atoms does not differ from the case of the free atom, shown in Fig. 1.1*a*.

1.3. Systematics of characteristic lines

1.3.1. The Moseley law

The relationship, reflecting the relation of the wavelength of the characteristic x-ray radiation with the atomic number of the element emitting the radiation represents the basis of x-ray spectrum analysis. This comparatively simple relationship was determined by Moseley [40] as a result of carrying out a large number of investigations into the systematisation of the experimental data on the characteristic x-ray spectra of a large number of elements.

At the same time, the Moseley law follows directly from the expression for the energy of the x-ray levels of the atom with a number of simplifying assumptions made in the expression. Actually, retaining in (1.2) only the main term of energy and omitting the multiplier $M_z/(M_z+m)$, which is close to unity, we obtain the following approximate expression for the energy of the level:

$$E = Rhc\frac{(Z-\sigma_1)^2}{n^2}.$$

The photon energy is determined by the difference of the energies of the initial i and final f levels of the atom. Therefore, for the energy of the x-ray photon:

$$\varepsilon = E_i - E_f = Rhc\left(\frac{(Z-\sigma_{1i})^2}{n_i^2} - \frac{(Z-\sigma_{1f})^2}{n_f^2}\right).$$

Selecting some mean full screening constants σ for two levels which does not disrupt the given equality gives;

$$\varepsilon = Rhc(Z-\sigma)^2\left(\frac{1}{n_i^2} - \frac{1}{n_f^2}\right),$$

or

$$\sqrt{\varepsilon} = (Z-\sigma)\sqrt{Rhc\left(\frac{1}{n_i^2} - \frac{1}{n_f^2}\right)}.$$

Taking into account that the photon energy is associated with its wavelength by the relationship:

$$\varepsilon = \frac{hc}{\lambda},$$

the resultant equation may be written in the form:

$$\frac{1}{\sqrt{\lambda}} = \chi(Z-\sigma), \tag{1.3}$$

where

$$\chi = \sqrt{\left(\frac{1}{n_i^2} - \frac{1}{n_f^2}\right)R}.$$

Equation (1.3) is the Moseley law for the lines of the x-ray spectrum: the quantity reciprocal to the square root of the wavelength of the characteristic radiation of the element depends in a linear manner on the atomic number of this element.

Since a number of assumptions was made in deriving equation (1.3), the Moseley law is approximate.

1.3.2. Spin-doublets in the x-ray spectrum

A spin doublet is two levels with the total quantum numbers n and l but differing in the direction of the spins, i.e. in the value of the internal quantum number j. Using equation (1.2) for the energy of the levels, it is easy to determine the distance on the energy scale between the levels of the spin-doublet:

$$\Delta E = Rhc\alpha^2 \frac{(Z-\sigma_2)^4}{n^3 l(l+1)}. \tag{1.4}$$

Equation (1.4) determines the difference of the energies of the spin levels.

The spin double lines are two lines having one initial transition level and two final levels which are the spin-doublet (or one final level and two initial levels, being the spin doublet).

Since one of the levels of the spin doublet lines with the energy E_0 is common, the energy of the first line is $\varepsilon_1 = E_1 - E_0$; the energy of the second line is $\varepsilon_2 = E_2 - E_0$.

It will be shown that the difference of the wavelengths of the spin-doublet lines depends only slightly on the atomic number of the elements. Actually, since the energy of the quantum is associated with its wavelength by the relationship:

$$\varepsilon = \frac{hc}{\lambda},$$

then

$$\Delta\varepsilon = -\frac{hc}{\lambda^2}\Delta\lambda.$$

Substituting the resultant value of the energy difference into equation (1.4) and solving the resultant expression in relation to $\Delta\lambda$, we obtain:

$$|\Delta\lambda| = R\alpha^2\lambda^2 \frac{\left(Z-\sigma^2\right)}{n^3 l\left(l+1\right)}.$$

Using the Moseley law, it is easy to obtain the following equation for λ^2:

$$\lambda^2 = \frac{1}{\chi^4 (Z-\sigma)^4},$$

or, substituting this expression into the previous relationship for $|\Delta\lambda|$, we obtain:

$$|\Delta\lambda| = \frac{R\alpha^2 (Z-\sigma_2)^4}{n^3 l(l+1)\chi^4 (Z-\sigma)^4}.$$

Assuming that σ_2 does not differ greatly from σ (this is close to the true value for the elements with the high atomic numbers), we finally obtain:

$$\Delta\lambda \approx \frac{R\alpha^2}{\chi^4 n^3 l(l+1)}. \tag{1.5}$$

Table 1.1. The difference of the wavelength of the spin doublet $K\alpha_{1,2}$ (in nm) for several Z

Z	13 (Al)	20 (Ca)	28 (Ni)	37 (Rb)	92 (U)
$\Delta\lambda$	$2.4\cdot10^{-4}$	$3.3\cdot10^{-4}$	$3.8\cdot10^{-4}$	$4.0\cdot10^{-4}$	$5.0\cdot10^{-4}$

Thus, it may be asserted that the difference of the wavelengths of the x-ray spin-doublet lines in the first approximation does not depend on the atomic number of the element. Table 1.1 gives the values of $\Delta\lambda$ for the lines $K\alpha_1$ and $K\alpha_2$ (the spin-doublet levels L_2 and L_3) for several atomic numbers Z.

Table 1.1 shows that the value of $\Delta\lambda$ greatly differs only at low values of Z for which the assumption on the equality $\sigma_1 = \sigma_2$ is very approximate. For the atomic numbers from $Z = 28$ to 92 the value $\Delta\lambda$ changes in the range from $0.38\cdot10^{-3}$ nm to $0.50\cdot10^{-3}$ nm, gradually increasing with the increase of the atomic number. The $K\alpha_{1,2}$ doublet is so narrow that, for example, in standard equipment with wave dispersion, designed for x-ray spectrum analysis, the doublet is usually not resolved. Equation (1.5) shows that the increase of the azimuthal quantum number l results in the formation of even more closely spaced spin-doublet lines.

1.3.3. Screening doublets

If two levels of the atom are characterised by the same quantum numbers n or j, but differ in the value of the azimuthal quantum number l, then in calculations of the energy of these levels using equation (1.2) we do not obtain values that coincide because the variation of the quantum number l is accompanied by a change of the constants of total and internal screening σ_1 and σ_2. The investigated pairs include the levels (L_1 and L_2), (M_1 and M_2), (M_3 and M_4), (N_1 and N_2), (N_3 and N_4), (N_5 and N_6), etc. These pairs of the levels are referred to as 'screening doublets'. The pair of the lines of the x-ray spectrum, formed in transition from each of these levels to one common level, or from one level to each level of such a doublet is also referred to as the screening doublet.

It will be shown that the difference of the values $\sqrt{E}$ for two lines of the screening doublet does not depend on the atomic number Z. Actually, in equation (1.2) the second term is considerably smaller than the first one because the constant of the fine structure α^2 is very small. Therefore, ignoring the second term and accepting that, as previously, $M_z/(M_z+m)\approx1$, we obtain

$$E = Rhc\frac{(Z-\sigma_1)^2}{n^2},$$

or

$$\sqrt{E} = \sqrt{Rhc}\frac{(Z-\sigma_1)}{n}.$$

The identical equation may be written for another level of the screening doublet:

$$\sqrt{E'} = \sqrt{Rhc}\frac{(Z-\sigma_1')}{n}.$$

For the difference of the resultant equations we obtain:

$$\Delta\sqrt{E} = \sqrt{Rhc}\frac{\Delta\sigma_1}{n}. \tag{1.6}$$

Thus, with a certain approximation, the difference of the square root of the energy of the levels of the screening doublets does not depend on the atomic number Z.

1.3.4. Determination of the screening constants σ_1 and σ_2

In order to use equation (1.2) for calculating the energy of the x-ray levels, it is necessary to know the values of the screening constants.

The method of determination of the internal screening constant σ_2 for the spin-doublet levels is quite simple. Actually, the formula for the difference of the energies of the spin-doublet lines, as mentioned previously, has the form:

$$\Delta E = Rhc\alpha^2\frac{(Z-\sigma_2)^4}{n^3 l(l+1)}.$$

The difference of the energies of the two levels ΔE may be measured. The remaining constants are available. Therefore, equation (1.4) makes it possible to determine the internal screening constant σ_2. The average values of σ_2 determined by this procedure for several levels are presented in Table 1.2 [7].

The calculation of the total screening constants σ_1 for the same spin-double of levels was carried out using the general equation (1.2) for the energy of the level and the internal screening constants σ_2, determined by the procedure described deviously. This is carried out using the experimental data on the energy of the levels. It appears that for the $L_{2,3}$-level, the constant σ_1 changes with the increase of the atomic number from 15 (for $Z = 45$) to 22 (for $Z = 92$), for the $M_{2,3}$-level from 27 to 41 with the value

Table 1.2. Values of the internal screening constant σ_2 for several spin-doublet levels

Level	$L_{2.3}$	$M_{2.3}$	$M_{4.5}$	$N_{2.3}$	$N_{4.5}$	$N_{6.7}$
n	2	3	3	4	4	4
l	1	1	2	1	2	3
σ_2	3.50	8.5	13.0	17.0	24	34

Table 1.3. Internal screening constants for single levels

Level	L_1	M_1	N_1
n	2	3	4
l	0	0	0
σ_2	2.0	6.8	14

of Z changing in the same range [7].

For the K-level $\sigma_1 = 0.3$ and depends only slightly on the atomic number Z.

For single levels (L_1, M_1, etc), the values of σ_1 and σ_2 cannot be determined by this procedure. Therefore, here we use the previously mentioned equation (1.6), which links the difference of the energies of the 'screening doublets' with the variation of $\Delta\sigma_1$ by the total screening constant. Calculating the value of $\Delta\sigma_1$ using the data on the energies of the levels of the screening doublets, we may use, for example, the volume σ_1 for the level $L_{2,3}$ (Table 1.2) and determine from (1.2) the value for the level L_1. The values of σ_1 for the single levels are slightly lower than σ_1 for the appropriate spin-doublets and change with the change of Z in approximately the same range as for the spin-doublets.

The internal screening constants σ_2 for the single levels are determined for the known values of the total screening constants σ_1 for these levels using equation (1.2) for the energy of the level and the experimental data for this energy. The values of the internal screening constants σ_2 for L_1-, M_1- and N_1-levels are presented in Table 1.3 [7].

1.3.5. The fine structure of the x-ray levels

The previously described equation (1.2) for the energy of the level is approximate. The complete equation contains another term which takes into account the angle between the vectors l and s in the magnetic interaction between the azimuthal and spin moments. Since the possible orientations in the space of the vector l are quantised, it is also necessary to introduce the magnetic quantum number m_l. The values of the magnetic quantum numbers are equal to $-l, -(l-1), -(l-2), ..., 0, ..., l-2, l-1, l$ for every l.

According to the Pauli exclusion principle, for the given atom the orbit characterised by the set of the quantum numbers n, l, m_l and m_s may contain only one electron. This means that in the atom there cannot be two electrons in the same energy state. Therefore, every energy level, characterised by the previously accepted quantum numbers n, l and s, breaks down into several individual similar sub-levels, forming the fine structure of the level.

Taking into account the existing set of the orientations of the azimuthal quantum moment, at the given value l there may be $2l+1$ states of the electrons. If it is also taken into account that every such state has two possible spins s, then for the given azimuthal quantum number l there maybe $2(2l+1)$ states of the electrons. Taking these considerations into account, Table 1.4 shows the configuration of the electrons on the K- and L-levels of the atom.

Similarly, for the M-shell we have:

M_1 – 2 electrons ($3s^2$);

M_2 – 2 electrons and M_3 – 4 electrons ($3p^6$);

M_4 – 4 electrons and M_5 – 6 electrons ($3d^{10}$).

Table 1.4. Distribution of the electrons in the atom with respect to the energy states on K- and L-levels

Level	n	l	m_l	m_s	g (number of electrons)	State
K	1	0	0	–1/2	2	$1s^2$
	1	0	0	+1/2		
L_1	2	0	0	–1/2	2	$2s^2$
	2	0	0	+1/2		
L_2	2	1	0	–1/2	2	$2p^2$
	2	1	0	+1/2		
L_3	2	1	-1	–1/2	4	$2p^4$
	2	1	-1	+1/2		
	2	1	+1	–1/2		
	2	1	+1	+1/2		

Table 1.5. Distribution of electrons in the states in the atoms of inert gases

2	He	$1s^2$										
10	Ne	$1s^2$	$2s^2$	$2p^6$								
18	Ar	$1s^2$	$2s^2$	$2p^6$	$3s^2$	$3p^6$						
36	Kr	$1s^2$	$2s^2$	$2p^6$	$3s^2$	$3p^6$	$3d^{10}$	$4s^2$	$4p^6$			
54	Xe	$1s^2$	$2s^2$	$2p^6$	$3s^2$	$3p^6$	$3d^{10}$	$4s^2$	$4p^6$	$4d^{10}$	$5s^2$	$5p^6$

In a filled shell with the main quantum number n, the total number of the electrons is $\sum_{l=0}^{n-1} 2(2l+1) = 2n^2$. This means that the filled K-shell (n=1) contains two electrons, in the L-shell (n = 2) there are eight electrons, and in the M-shell (n=3) there are 18 electrons, and so on.

Table 1.5 shows the distribution of the electrons in the atoms of inert gases. The table shows that in the xenon atoms (54 Xe) there are no electrons in the f-state of the penultimate shell. This disruption in the filling of the shell is caused by the fact that the energy of the s-state of the next shell is lower than the energy of the f-state of the previous shell ($4f$-shell is filled in rare-earth elements).

1.4. Intensity of the lines of the characteristic spectrun

The intensity N_{ij} of the line of characteristic radiation is determined by the expression:

$$N_{ij} = N_1 F_i \omega_i p_{ij} F_j, \tag{1.7}$$

where N_1 is the intensity of the exciting corpuscular or radiation flux falling on the specimen, F_i is the probability of ionisation by this flux of the internal levels i of the atoms included in the composition of the emitting material, ω_i is the probability of appearance of x-ray radiation after ionisation of the level i (fluorescence yield), p_{ij} is the probability of the radiation transition of the atom from the state i to state j, F_j is the probability of emission of the resultant radiation from the specimen in the given direction taking into account its absorption in the specimen.

In equation (1.7) the intensity of the primary flux N_{ij} is represented by the number of photons (or particles) per unit time on the unit area, normal to the direction of the beam. To convert the value of N_{ij} to intensity I_{ij}, expressed in energy units, equation (1.7) must be multiplied by the energy of the formed photon $h\nu_{ij}$:

$$I_{ij} = N_1 F_i \omega_i p_{ij} F_j h\nu_{ij}. \tag{1.7a}$$

1.4.1. The relative intensity of lines

Taking into account equation (1.7) and assuming that $F_j = 1$, the following equation is obtained for the relative intensity of the lines belonging to a specific series of the x-ray spectrum:

$$\frac{N_{i1,j}}{N_{i2,f}} = \frac{F_{i1} p_{i1,j}}{F_{i2} p_{i2,f}}, \tag{1.8}$$

where the indexes $i1$ and $i2$ characterise the sublevels of the i-shell (initial state of the atom), and the indexes j and f characterise the sublevels of the final states of the atom.

In equation (1.8), the probabilities F_{i1} and F_{i2} of the ionisation of the sublevels $i1$ and $i2$ may be regarded as proportional to the number of electrons g_{i1} and g_{i2} on these sublevels. Therefore

$$\frac{N_{i1,j}}{N_{i2,f}} = \frac{g_{i1} p_{i1,j}}{g_{i2} p_{i2,f}}. \tag{1.9}$$

1.4.1.1. Selection rules

The relative intensity of the lines of the x-ray spectrum is determined primarily by the sampling rules of quantum mechanics according to which a number of transitions between the levels i and f are characterised by a very low probability and the value of $p_{i,f}$ for these transitions in equation (1.9) may be regarded as equal to 0. The highest probability of appearance is characteristic of the 'dipole' lines determined by transitions with the following variation of the quantum numbers:

$$|\Delta l| = 1, \ |\Delta j| = 0,1; \quad \Delta n \neq 0.$$

The 'dipole' lines of the K-series with the highest intensity are $K\alpha_1$, $K\alpha_2$, $K\beta_1$, $K\beta_2$ and $K\beta_3$ lines, shown in Fig. 1.2. At the same time, there are no lines associated with the transitions $K \rightarrow L_1$, $K \rightarrow M_1$, and $K \rightarrow N_1$, because in this case $\Delta l = 0$.

The x-ray spectrum also contains very weak lines corresponding to some 'forbidden' transitions. They include the 'quadrupolar' lines for which

$$|\Delta l| = 0,2 \ \text{ and } \ |\Delta j| = 0, 1, 2.$$

The intensity of the 'quadrupolar' lines is approximately two orders of magnitude smaller than the intensity of the 'dipole' lines.

In addition, there are 'non-diagram' lines which generally cannot be

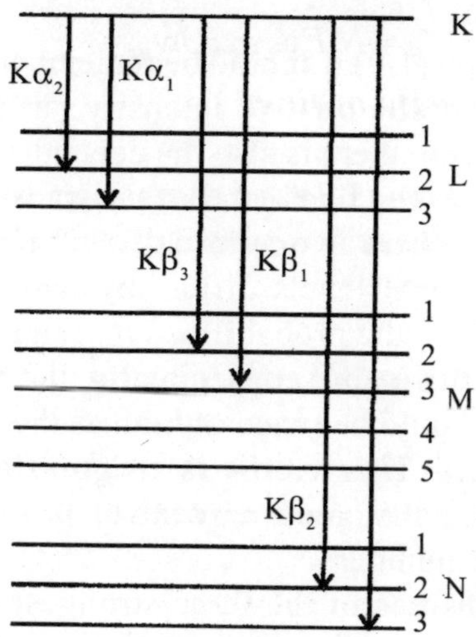

Fig. 1.2. The main dipole lines of the x-ray K-spectrum.

identified using the conventional scheme of the energy levels of the atom.

For example, the Kα-linc has several satellites on the shortwave side. The intensity of these satellites amounts to whole percent of the intensity of the Kα-peak. These satellites are also found in the case of the $K\beta_1$-line and in the lines of the L- and M- series. They are determined by the transitions in the case of twice-ionised atoms. Their displacement in the direction of the shortwave side is explained by the fact that the additional vacancy results in a decrease of the efficiency of screening the charge of the nucleus and, consequently, in an increase of the bonding strength of the electrons in the atom.

1.4.1.2. Effect of the 'population' of the levels on the relative intensity of multiplet lines

The second factor as regards importance (after the selection rules) which determines the relative intensity of the lines is the factor of 'population' by electrons of the levels (sub-levels) between which transitions take place.

A suitable example of the effect of 'population' is the ratio of the intensity of the lines, detected in a multiplet.

The multiplet is a group of lines for which the initial conditions have the same quantum numbers *n* and *l* but they may differ in the direction of spins, and the final states have the common quantum numbers *n'*, *l'* and may also differ only in the spin direction. The multiplets include, for example, the doublets ($K\alpha_1$- and $K\alpha_2$-lines) and (Lη- and L*l*-lines), and triplets ($L\alpha_1$-,

$L\alpha_2$- and $L\beta_1$-lines).

Examining the equation (1.9) it should be remembered that in addition to the explicit dependence of the relative intensity on 'population' g_i of the initial level of the transition, there is also the dependence of the probability p_{ij} on the 'population' g_j of the final level of the transition.

For the doublet $K\alpha_{1,2}$ there is one initial level K. Therefore, $g_{i1} = g_{i2}$. The 'population' of the final levels differs by two orders of magnitude ($g_{L3} = 4$; $g_{L2} = 2$). Therefore, the probabilities $p_{i1,j}$ and $p_{i2,f}$ in equation (1.9) also differ by a factor of two and, consequently, the intensity of the $K\alpha_1$-line (the final level L_3) should be twice as high as the intensity of the $K\alpha_2$-line (the final level L_2). This result is in good agreement with the experimental data. Especially good agreement is found in the range of medium and high atomic numbers.

The ratio of the intensities of the three strongest lines of the L-series (triplet $L\alpha_1$, $L\alpha_2$ and $L\beta_1$) may be evaluated using a simplified approach based on the Burger–Dorgello rule for multiplets (Fig. 1.3). Using this rule, it is initially assumed that the levels L_2 and L_3 merge into a single level (case A). Consequently, from the ratio of the levels M_4 and M_5 we obtain the ratio of the intensities:

$$\frac{N L\alpha_1}{N L\alpha_2 + N L\beta_1} = \frac{6}{4}.$$

It is then assumed that M_4 and M_5 merge into a single level but the levels L_2 and L_3 are separated (case B). Consequently, from the population of the levels L_2 and L_3 we obtain the ratio of the intensities

$$\frac{N L\alpha_1 + N L\alpha_2}{N L\beta_1} = \frac{4}{2}.$$

Solving jointly the resultant two equations, and assuming that $IL\alpha_2=1$, we obtain

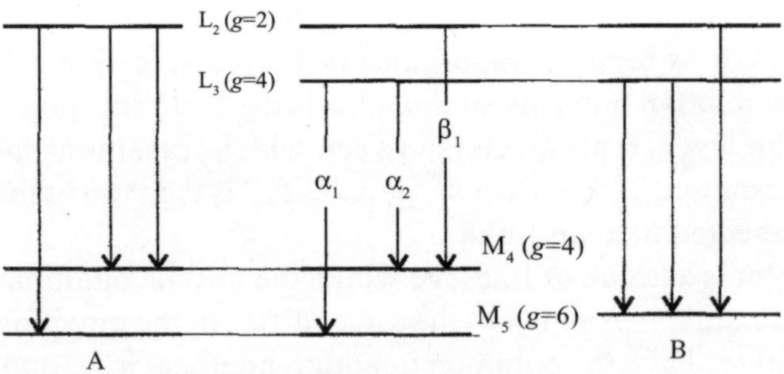

Fig. 1.3. Example of the determination of the relative intensities of the main lines of the L-series using the Burger–Dorgello rule.

$$NL\alpha_1 : NL\alpha_2 : NL\beta_1 = 9:1:5.$$

The resultant dependence determines the ratio of the intensities, expressed by the number of quanta of x-ray radiation of the component lines. To transfer to the intensities represented by energies, it is necessary in accordance with equation (1.7a) to take into account the difference of the quantum energy (or wavelength) of the investigated radiation. Taking these considerations into account and accepting that the intensity of the $L\alpha_1$-line is 100 (as is usually assumed in the handbook literature), we obtain:

$$IL\alpha_1 : IL\alpha_2 : IL\beta_1 = 100:11:59.$$

This relationship is in quite a good agreement with the experimental data. For example, according to the data in [41], in the case of silver this ratio is 100:11:71, and for tantalum 100:12:47.

The resultant ratio is slightly different for thick specimens because it is affected by the differences in the absorption of the component lines in the matter. In particular, the relative intensity of the shortwave $L\beta_1$-line greatly increases.

'The population' of the non-filled outer levels of the atom increases with the increase of the atomic number. This determines the probability of the transition p_{ij} and, correspondingly, the intensity of the lines associated with the transition of the atoms to the investigated state. For example, in the free atom the $K\beta_{1,3}$-line appears only for the element with the atomic number $Z = 13$ (Al), when the filling of the $M_{2,3}$ sublevel starts ($K\beta_{1,3}$-line is determined by the transition K–$M_{2,3}$). Subsequently, with the increase of the atomic number from 13 to 18 (and with the increase of the number of the electrons on the $M_{2,3}$-sublevel) there is a monotonic increase of the intensity of the given line and, consequently, the ratio of the intensity of the $K\beta_{1,3}$-line to the intensity of the $K\alpha$-line increases.

It should be mentioned that in a solid body, the line in the spectrum of some element forms prior to the moment when the electron appears on the final level of the transition of the free atom. For example, the previously examined $K\beta_{1,3}$-line appears in the solid for the first time not in the case of aluminium ($Z = 13$) but for sodium ($Z = 11$), although according to the handbook data for free electrons, the generating level $M_{2,3}$, corresponding to the $K\beta_{1,3}$-line in the case of sodium does not contain electrons. This effect is explained by the influence of the chemical bonds which transferred the outer electrons to other optical levels from which they may carry out permitted transitions to the vacancies of the K-level.

Table 1.6 gives the elements for which the first electrons appear in the free atoms on the generating levels, and the elements in which the corresponding line forms in the solid for the first time. In all investigated cases, the initial level is the K-level.

Table 1.6. Atomic numbers Z of elements for which the first lines of the K-series appear in the solid and in the free atom

Level	K-series line	Solid	Free atom
$L_{2.3}$	$\alpha_{1.2}$	3 Li	5 B
$M_{2.3}$	$\beta_{1.3}$	11 Na	13 Al
$M_{4.5}$	β_5	17 Cl	21 Sc
$N_{2.3}$	β_2	30 Zn	31 Ga

Table 1.6 shows that all the lines of the K-series in the solid form prior to the appearance of the first electron on the corresponding level of the free atom.

1.4.1.3. Effect of the Koster-Kronig transition on the relative intensity of lines

The next factor, affecting the relative intensity of the lines, is the redistribution of the vacancies between the sublevels of a single shell.

It was mentioned previously that the radiation transitions inside the same series ($\Delta n = 0$) are forbidden by the selection rules. However, the non-radiation transitions (or Auger transitions) with the injection of the electron from some outer shell are fully possible in this case.

These transitions were detected for the first time by D. Koster ad R. Kronig [42] in 1935. This type of Auger transition was subsequently named after them.

Regrouping of the vacancies inside a shell takes place prior to the completion of the radiation transition between the shells. Therefore, the relative intensities of the lines (all the series with the exception of the K-series), associated with the filling of these vacancies, depend on the probability of the Koster–Kronig transitions.

The redistribution of the vacancies will be examined on the example of the L-shell. The initial distribution of the vacancies, determined by some process of ionisation of the shell, corresponds in the first approximation to the 'population' g of the sublevels of the L-shell. This means that:

$$g_{L1} : g_{L2} : g_{L3} = 1:1:2.$$

As a result of the Koster–Kroning transitions, this ratio of the vacancies should change. f_{12} denotes the probability of the transition of a vacancy from the L_1- to L_2-sublevel ($L_1 \rightarrow L_2$); f_{13} is the probability of the $L_1 \rightarrow L_3$

transition, and f_{23} the probability of the $L_2 \to L_3$ transition. Then, the new vacancies V are redistributed in the L-series as follows:

$$V_{L1} : V_{L2} : V_{L3},$$

where

$$V_{L1} = g_{L1}(1 - f_{12} - f_{13}),$$
$$V_{L2} = g_{L2}(1 - f_{23}) + g_{L1} f_{12},$$
$$V_{L3} = g_{L3} + g_{L1} f_{13} + g_{L2} f_{23}.$$

This means that the vacancies are displaced to the sublevels L_2 and L_3 with a lower binding energy but the total number of the vacancies naturally remains unchanged. The increase of the number of the vacancies on the L_3-sublevel increases the intensity of the $L\alpha_1$- and $L\alpha_2$-lines because these strong lines of the L-series are associated with the transitions of the electrons to the L_3-sublevel (Fig. 1.3).

Attention will now be given to the probabilities f_{12}, f_{13} and f_{23} of the Koster–Kronig transitions inside the L-shell. These transitions are accompanied by the ejection of the Auger electrons from the shells away from the atomic nucleus.

The case in which donors of the Auger electrons are the M_4 and M_5 sublevels of the M-shell, was investigated in [7]. For the Koster–Kronig transitions to take place, the difference of the energies of the L-sublevels between which the transition takes place should exceed the binding energy of the electrons on the M_4- or M_5-sublevels. It should be taken into account that the absence of the electron on the L-shell in the ionised atom increases the binding strength of the electrons of the M-shell with the atomic nucleus. In the first approximation this intensification may be described numerically as the increase of the charge of the nucleus by unity (or the increase by unity of its atomic number), i.e.:

$$E_{L1}(Z) - E_{L3}(Z) > E_{M4,M5}(Z+1),$$
$$E_{L2}(Z) - E_{L3}(Z) > E_{M4,M5}(Z+1). \qquad (1.10)$$

These conditions are not fulfilled for all elements. Figure 1.4 compares the dependence of the difference of the energies of the L_1- and L_3-sublevels on the atomic number and the dependence on this parameter of the binding energy of the electrons of the M_5-sublevel.

Figure 1.4 shows that the $L_1 \to L_3M_5$ transition is possible only for the elements with the atomic number $50 > Z > 73$. Similarly, the transition $L_1 \to L_3M_4$ is possible for $50 > Z > 77$, and the transition $L_2 \to L_3M_{4.5}$ is possible only for $Z < 30$. Naturally, the probabilities $f_{1,3}$ and $f_{2,3}$ are the highest for the same atomic numbers of the elements. If the difference of the energies

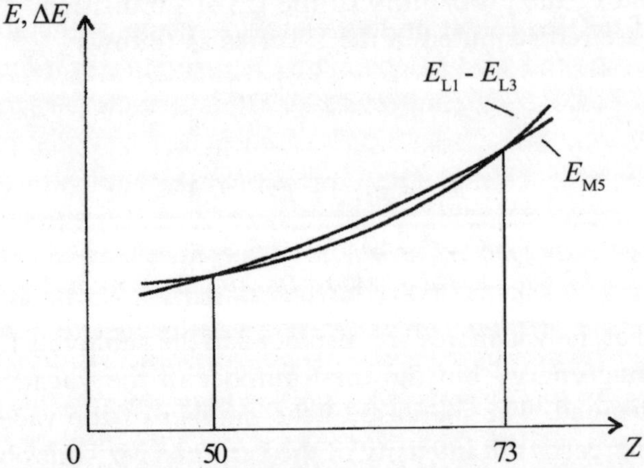

Fig. 1.4. The ratio of the differences of the energies of the two L-sublevels and the energy of the M_5-level as the function of the atomic number of the element.

of the L-subshells is lower than the binding energy of the $M_{4,5}$-level, than the appropriate probabilities of the Koster–Kroning transitions are associated with the ejection of the Auger electrons from the shells more remote from the nucleus N, O, etc.

The occurrence of the $L_1 \rightarrow L_3M_{4,5}$ transitions is confirmed experimentally by the large increase of the intensity of the shortwave satellites of the $L\alpha_1$- and $L\alpha_2$-lines (determined by double ionisation) for the elements whose atomic numbers correspond to the previously mentioned conditions.

1.4.1.4. Determination of the probability of intra-atomic transitions p_{if}

The probability of the transitions p_{if} of the atom from the state with the ionised level *i* to the state with the ionised level *f* determines, as mentioned previously, the relative intensity of the lines of the characteristic radiation inside some series (K-, L-, etc.). The calculation of these quantities is a complicated quantum mechanics task. Therefore, in practice, the values of p_{if} are usually determined using the experimental data on the relative intensity of the lines of some series of x-ray radiation. The tabulated values of the relative intensities are given as the ratio to some high-intensity diagram line of the given series. In the case of the K-series such line is usually $K\alpha_1$ whose intensity is regarded as 100%. The probability p_{if} may be calculated in relation to all the lines of the series or in relation to the group of lines belonging to the sublevel to which the given line relates:

$$p_{i,f} = \frac{I_{i,f}}{\sum_f I_{i,f}}. \tag{1.11}$$

The possibility of using this formula is explained by the fact that after ionisation of the atom, the processes taking place in the atom are intra-atomic processes and do not depend on the external conditions. Therefore, for the given sublevel, the investigated ratio characterises the probability of the intra-atomic transitions. The K-shell does not have any sublevel, and the formula (1.11) holds for the entire K-series. The probabilities of excitation of different sublevels for the L-, M- and other shells differ and depend on the conditions and method of ionisation. Therefore, calculations using equation (1.11) of the probability p_{if} are usually carried out in relation to the sum of the lines for every sublevel, regarding it as a separate level.

Reports on the calculated and experimental relative intensities of the characteristic lines of the x-ray spectrum may be found in [43–45].

1.4.2. X-ray fluorescence yield ω_i

The ionised atom may return to the non-excited state by two methods: by the emission of the x-ray photon (radiation transition) or by ejection of the electron from the outer shell (radiation less transition or Auger transition).

The probability ω_i of the radiation transition, used in expression (1.7), is also referred to as the x-ray fluorescence yield and is the property of the given atom and depends only on the atomic number Z (an exception are the levels close to the outer shells of the atom where the effect of chemical bonds is strong. For example, according to [46], the value of ω_K for oxygen in O_2 is equal to 0.0073, in CO_2 it is 0.0094, and in H_2O it is 0.0131). This probability may be regarded as the ratio of the number of the atoms of the element, emitting the x-ray quantum, to thc total number of the ionised atoms. If the probability of the Auger transitions is denoted by a_i, then $\omega_i + a_i = 1$. This means that the filling of the vacancy on the i-level is possible only by these two processes which mutually exclude each other.

After filling the vacancy on the i-level, the atom is in the state of single (radiation transition) or double (Auger transition) ionisation. The return of the atom to the non-excited state is accompanied by a cascade of successive transitions.

The x-ray fluorescence yield ω_q for the q-shell (specifically for K- and L-shells) of the atoms with different atomic numbers Z of the element, is shown in Fig. 1.5.

The dependences, shown in Fig. 1.5, are approximated quite efficiently by the formula proposed by Wentzel [47]:

$$\varpi_q = \frac{Z^4}{C_q + Z^4}, \tag{1.12}$$

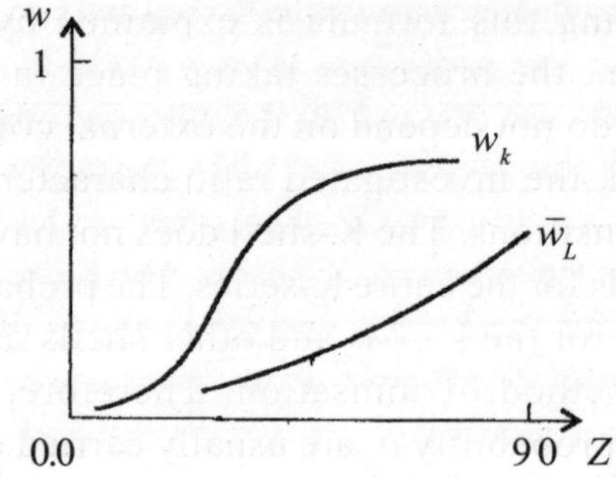

Fig. 1.5. X-ray fluorescence yield as a function of the atomic number Z.

Here $C_K = 1.06 \cdot 10^6$, $C_L = 1.0 \cdot 10^8$, $C_M = 1.4 \cdot 10^9$.

Figure 1.5 and equation (1.12) show that the x-ray fluorescence yield is very small for the elements with low atomic numbers. For example, for sodium ($Z = 11$) it is only approximately 1%. For high atomic numbers Z, the value of ω_K exceeds 90%. The values of the fluorescence yields, averaged with respect to the data from 11 sources, are presented in [33].

Formula (1.12) has been improved many times by a number of authors (see, for example [48–50]). The polynomial model, presented in [50] has the following form:

$$\left(\frac{\omega_q}{1-\omega_q}\right)^{1/4} = C_0 + C_1 Z + C_2 Z^2 + C_3 Z^4,$$

where according to [51] the coefficients for the K-shell are determined as: $C_0 = 0.0370$, $C_1 = 0.03112$, $C_2 = 5.44 \cdot 10^{-5}$, $C_3 = -1.25 \cdot 10^{-6}$. For the L-shell (at $Z > 30$) these coefficients have the values proposed in [52]: $C_0 = 0.177650$, $C_1 = 0.00298937$, $C_2 = 8.91297 \cdot 10^{-5}$, $C_3 = 2.67184 \cdot 10^{-7}$.

It should be mentioned that the investigated formulae given the mean values of the x-ray fluorescence yield for the entire q-shell. In fact, the intensity of the characteristic lines for the L-, M- and other shells, is determined by the transitions of the electrons to specific sublevels of these shells. Therefore, it is more accurate to talk about the x-ray fluorescence yields ω_{q1}, ω_{q2}, ω_{q3} for the individual sublevels. The values of these yields are presented in Table 1.7.

In the calculations of the intensity of the x-ray lines, the selection of the fluorescence yield ω_i for L-, M-, etc. shows should be matched with the value of the probability of transition p_{if}. If ω_i is selected for the sublevel, then the probability p_{if} is examined with respect to the transitions of these sublevel. If the value ω_i, determined by the equation (1.12) and averaged out with respect to the shell is taken into account, p_{if} should be calculated for the entire shell.

Table 1.7. The yield of x-ray fluorescence for the K-shell and several sublevels of L- and M-shells

Element	ω_K	ω_{L1}	ω_{L2}	ω_{L3}	ω_{M5}
Al (13)	0.0349	0.000026	0.00075	0.00075	–
Cu (29)	0.437	0.0026	0.010	0.011	–
Ag (47)	0.827	0.016	0.051	0.052	–
W (74)	0.958	0.147	0.270	0.255	0.021
Pb (82)	0.971	–	–	0.359	0.031
U (92)	0.974	0.176	0.467	0.489	0.050

1.4.3. The intensity of characteristic radiation, excited by the electron flux

In examination in section 1.4.1 of the relative intensity of the lines of characteristic x-ray radiation, no attention was given to the method of ionisation of the atoms. At the same time, the probability F_i of the ionisation of the level *i* of the atoms of the specimen and the probability F_j of emission of x-ray radiation from the specimen, included in equation (1.7), depends greatly on the method of excitation of the atoms of the investigated material. The flux of charged particles is often used for these purposes. The case of excitation of atoms of matter by an electron flux will be investigated.

1.4.3.1. Basic equation for the intensity of characteristic radiation, excited by electrons

In the investigated case, probability F_i of ionisation of the level *i* is the ratio of the number of the atoms, ionised by the electron beam, to the number of these electrons in the beam. In other words, the value F_i is determined by the number *n* of ionisations of the atoms, related to a single incident electron. The number of ionisations *dn* in the elementary section of the path *dx* of the electron in matter is equal to the product of the ionisation cross-section *Q* by the number of atoms in the unit volume. For the single-element object we have:

$$dn = Q\frac{N\rho}{A}dx, \tag{1.13}$$

where $N\rho/A$ is the number of atoms in the unit volume, $N = 6.02 \cdot 10^{23}$ is the Avogadro number, A is the atomic weight of the element which forms matter, ρ is the density of matter.

The ionisation cross-section Q is the cross-section in which the probability of ionisation of the atom in collision with the electron is equal to 100%. A number of expressions has been proposed for the cross-section Q. The most suitable equation is the one proposed by Bethe [53] derived in 1930 on the basis of quantum mechanics considerations:

$$QE_q^2 = a\frac{\ln(rU)}{U}, \tag{1.14}$$

where E_q is the excitation potential of the q-shell of the atom, $U = E/E_q$ is the amount by which the energy of the incident electrons E exceeds the excitation potential E_q, r is a constant quantity which at $U < 20$ may be regarded as equal to 1 for the ionisation cross-section of the K-level.

The authors of [54, 55] proposed a modified Bethe equation:

$$QE_q^2 = a\frac{\ln U}{U^m}. \tag{1.14a}$$

In [54] it was assumed that for the K- and L-series $m = 0.9$, and for the M-series $m = 0.7$. In [55] it was accepted that for the K-series $m = 0.95$, and for the L- and M-series $m = 0.8$. Since the values of m have not been stabilised, in considerations of the ionisation cross-section we shall use the Bethe equation in the form (1.14).

Coefficient a in equation (1.14) was determined as $a = \pi e^4 b_q g_q$, g_q is the number of electrons in the q-shell of the atom ('population' of the shell), b_q is the ionisation parameter of the q-shell, determined in experiments.

The value of the parameter b_K is determined with sufficient accuracy for the K-shell. According to [18], $b_K = 0.35 \cdot 1.73 = 0.606$ and does not depend on the atomic number. In [56], the value of b_q for the K- and L-shells is examined as the function of the atomic number Z:

$$b_q = \frac{A_q}{Z - B_q} + C_q + D_q Z, \tag{1.15}$$

where the values of the approximation coefficients A_q, B_q, C_q, D_q for the K-series are selected equal to: $A_K = 1.1287$, $B_K = 9.25$, $C_K = 0.556$, $D_K = 0.00198$. For the L-series: $A_L = 2.519$, $B_L = 26.6$, $C_L = 0.0968$, $D_L = 0.0103$. Equality (1.15) is determined for high electron energies (150-200 keV) where the constant r in the equation (1.14) differs from unity and has the values $r_K = 0.80$ and $r_L = 0.95$. The values of parameter b_L, calculated from (1.15), are presented in Table 1.8.

In [56], the value of the parameter b_M in the ionisation section for the

Table 1.8. Dependence of parameter b_L on the atomic number of elements

Element	Rh (45)	La (57)	W (74)	Pb (82)
b_L	0.50	0.57	0.72	0.79

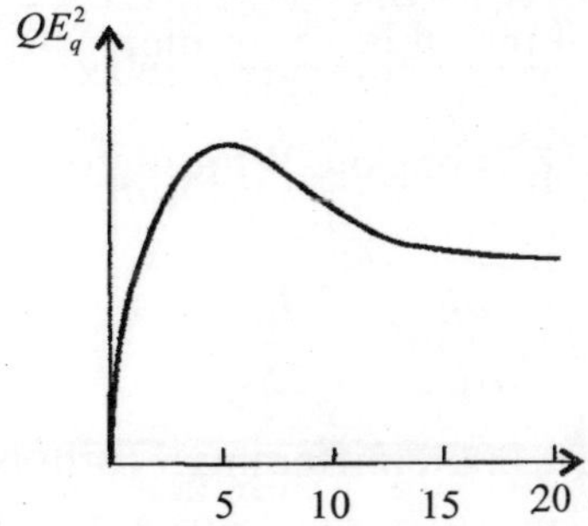

Fig. 1.6. Dependence of the ionisation cross-section on the amount by which the energy of the incident electron exceeds the binding energy of the electrons of the q-shell.

M-shell was assumed to be equal to $b_M = 0.75$. It was also assumed that $r_M = 0.60$. Further, it is assumed that $U < 20$ and, therefore, $r = 1$.

Equation (1.14) shows that the ionisation section depends strongly on the energy of the incident electron. The dependence of QE_q^2 on the 'overvoltage' U is shown in Fig. 1.6.

As regards the order of magnitude, the ionisation cross-section Q for the K-shell is equal to approximately 10^{-23} cm^2, i.e. less than 0.001 nm in diameter. If it is taken into account that the diameter of the atom is approximately 0.1 nm, it is clear that the probability of interaction of the electron with the atom is small. In other words, a very small part of the electrons causes ionisation of the K-shell of the atoms, prior to their complete deceleration. Braking of the electrons in matter takes place as a result of multiple inelastic collisions with the electrons of the outer shells of the atoms. Therefore, it may be assumed that braking is continuous. Consequently, to determine the number of ionisations along the entire braking path of the electron, equation (1.13) may be integrated over the length of the path:

$$F_i = n = \frac{N}{A} \int_x Q d(\rho x). \tag{1.16}$$

To reduce the subintegrand to a single variable, it is necessary to consider the dependence of the electron energy on the path travelled by the electron, i.e., the braking law of the electrons. A number of expressions have been proposed by different authors for this dependence. We examine two such expressions: the Thomson–Widdington law and the Bethe law.

The Thomson–Widdington law [57] for the dependence of the electron energy on the path travelled by the electrons is represented by a simple relationship:

$$E_0^2 - E^2 = b\rho x,$$

where E_0 is the initial electron energy, b is some quantity usually regarded as constant (generally speaking, this value increases with increasing E_0 and decreases with increasing Z).

In the differential form, the Thomson–Whittington law has the following form:

$$\frac{dE}{d(\rho x)} = -\frac{b}{2E}. \tag{1.17}$$

The more accurate Bethe law in the differential form is represented by the expression:

$$\frac{dE}{d(\rho x)} = \text{const} \cdot \ln\left(\frac{1.166\,E}{J}\right)\frac{Z}{A \cdot E}, \tag{1.18}$$

where $J = 11.5Z$ is the mean excitation energy of the atom. (This expression for J is the simplest expression. Other variants of the dependence of the mean excitation energy on the atomic number have been examined in, for example, monographs [23]). The application of the constant 1.166 in (1.18) assumes that energy E is taken in kilovolts.

The modified Bethe equation, proposed in [58], has the form:

$$\frac{dE}{d(\rho x)} = -78.5 \ln\left(0.85 + \frac{1.166 \cdot E}{J}\right)\frac{Z}{A \cdot E}.$$

If it is taken into account that $Z/A \approx 1/2$, the Bethe law confirms that the variation of the electron energy is inversely proportional to the double energy of the electrons, as indicated by the Thomson–Widdington law. The difference is that the Bethe law determines the dependence of the coefficient on the electron energy and the atomic number of matter.

Using the Thomson–Widdington formula (1.17) and also the cross-section Q, given by the expression (1.14), where the value of r is taken as equal to unity, from formula (1.16) we obtain:

$$F_i = \frac{2aN}{bA} \int_1^{U_0} \ln U dU.$$

Consequently, after integration

$$F_i = \frac{2aN}{bA}(U_0 \ln U_0 - U_0 + 1),$$

or, assuming that $(U_0 \ln U_0 - U_0 + 1) \approx 0.365\,(U_0-1)^{1.67}$, we obtain [18]:

$$F_i = \frac{0.730aN}{bA}(U_0 - 1)^{1.67}. \tag{1.19}$$

Substituting the value of F_i from (1.19) into (1.7) and ignoring the absorption of radiation in the specimen, we obtain:

$$N_i = \frac{0.730aN}{bA}(U_0 - 1)^{1.67}\,\omega_i p_{if}\left[\frac{\text{Photon}}{\text{electron}\times\text{s}\times\text{steradian}}\right]. \tag{1.20}$$

If we used the Bethe braking law, represented by the formula (1.18), we obtain a more complicated dependence for the intensity of the characteristic x-ray radiation:

$$N_i = \frac{g_q b_q}{4\pi}\frac{1}{ZE_q}\left(U_0 - 1 - \frac{\ln X}{X}\big(Li(XU_0) - Li(X)\big)\right)\omega_i p_{ij}, \tag{1.21}$$

where $X = \dfrac{1.166E_q}{J}$ and Li is the integral logarithmic function.

If we use the concept of the mean energy $\overline{E}$ of the braking electrons, expression (1.21) is simplified to the following form:

$$N_i = \frac{g_q \times b_q}{2\times 4\pi}\frac{1}{Z\ln X}(U_0 \ln U_0 - U_0 + 1)\,\omega_i p_{ij} \tag{1.22}$$

in units [photon/(electron·s·steradian)], where $X = \dfrac{1.166\overline{E}}{J}$.

1.4.3.2. Taking into account backscattering of the electrons

The basic equation (1.22), like the expressions (1.20) and (1.21) for the intensity of the characteristic spectrum, excited by the electron beam, does not take into account the process of backscattering of the electrons. As a result of multiple scattering, some of the electrons falling on the specimen leave the latter and do not take any part in the excitation of the atoms of the specimen. This process greatly reduces the probability F_i of ionisation of the atoms of matter.

The number of backscattered electrons increases with the increase of the atomic number of matter and, for example, for tungsten it reaches approximately half of the number of the electrons falling on the surface of the specimen. The energy distribution of the electrons with the initial energy of 30 keV, backscattered under the angle of 135°, determined by experiments carried out by Bishop [59], is shown in Fig. 1.7.

Figure 1.7 shows that with the increase of the atomic number of the target, the maximum of the energy distribution of the backscattered electrons

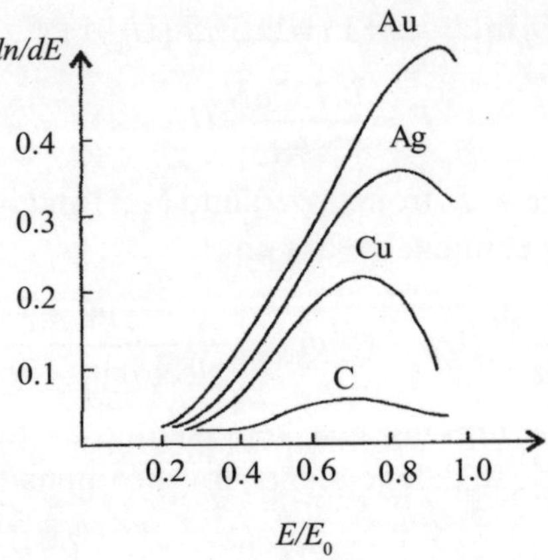

Fig. 1.7. Dependence of the energy distribution of the backscattered electrons on the material of the target ($\vartheta = 135°$, $E_0 = 30$ keV).

is displaced to the range of high energies. This means that for the high atomic numbers, backscattering takes place mainly in the subsurface layer, and the electrons which leave the specimen lose a small part of their energy.

The expression for the energy distribution of the backscattered electrons in relation to the atomic number of the target and the energy of the incident electrons has been presented in [60, 61]. For calculations in practice it is recommended to use the concept of the factor R of the backscattering of the electrons. This factor shows how many times the intensity of x-ray radiation decreases as a result of the effect of backscattering of the electrons. A correction of this effect is done by multiplying the expression for the intensity of characteristic radiation by the value R which is always smaller than unity. The dependence of the factor R of the backscattering of the electrons on the atomic number of the target and the energy of the scattered electrons, determined in [60], is presented in Fig. 1.8.

The factor of backscattering of the electrons in the calculations of the intensity of x-ray radiation is taken into account using very complicated semi-empirical expressions examined in, for example, [21–23].

1.4.3.3. Taking into account the absorption of characteristic x-ray radiation

The formulae (1.20)–(1.22) for the intensity of characteristic x-ray radiation do not take into account the absorption of this radiation emitted from the specimen.

Absorption may be taken into account by introducing a correction for absorption $f(\chi)$ determined by integration over the depth x of some function

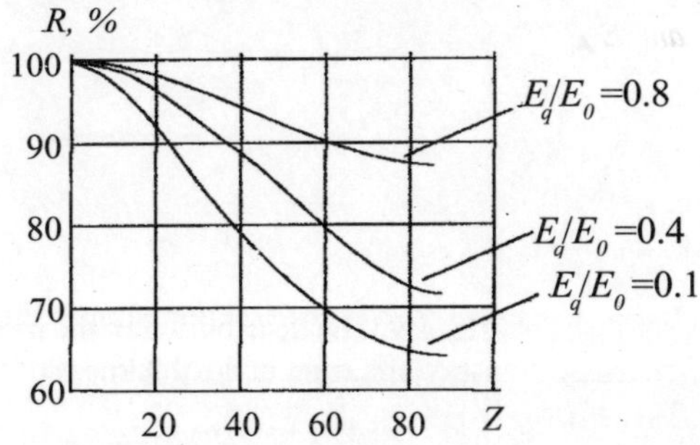

Fig. 1.8. Dependence of the coefficient of backscattering of the electrons on their energy and the atomic number of the target.

$\varphi(\rho x)$ which determines the distribution of the characteristic x-ray radiation generated in the specimen:

$$f(\chi) = \int_0^\infty \varphi(\rho x) \exp(-\chi\rho x) d(\rho x), \tag{1.23}$$

where $\exp(-\chi\rho x)$ takes into account the attenuation of emitted radiation:

$$\chi = \frac{\mu}{\sin\psi},$$

μ is the mass coefficient of attenuation of radiation in the specimen, ψ is the angle of exit of the radiation from the specimen.

The main difficulty in the evaluation of the effect of absorption on the intensity of characteristic x-ray radiation is the complicated procedure of determination of the type of function $\varphi(\rho x)$, included in equation (1.23). The complexity of this function is determined by the randomness of the distributions of deceleration of the electrons inside the specimen and by the strong dependence of the function on the atomic number of the target.

The search for the distribution $\varphi(\rho x)$ for characteristic radiation started a long time ago (see, for example [62]). One of the first studies in this direction was the experimental determination of $\varphi(\rho x)$, carried out by Castaing and Descamps [63]. At present, there are large banks of the experimental data for the investigated distribution containing hundreds of measurements (see, for example [64, 65]). Consequently, it is possible to ensure high accuracy of the approximation of the function $\varphi(\rho x)$. In a general case, the distribution $\varphi(\rho x)$ of characteristic x-ray radiation in the depth has the form shown in Fig. 1.9.

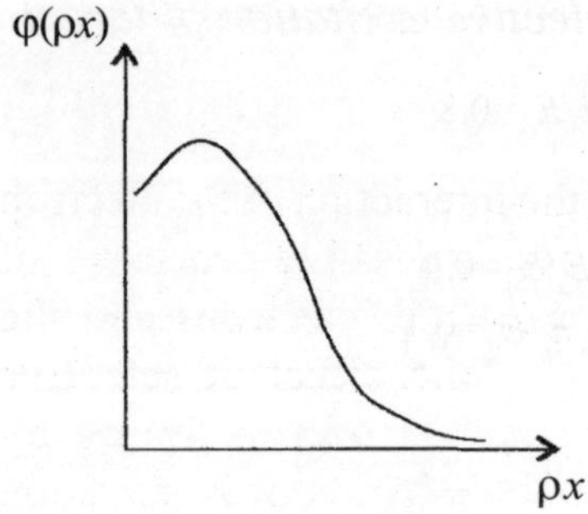

Fig. 1.9. The distribution of the primary characteristic x-ray radiation in the thickness of the emitting object.

A large number of functions is available for the approximation of the empirical relationships [23, 54, 66]. A detailed examination of these approximations [67] shows that almost any of them may be described by the exponential-power relationship:

$$A(\rho x + B)^m \exp\{C(\rho x + D)^n\},$$

where m and n are the integers which have the values in the range 0–2, and A, B, C and D are some parameters.

One of the comparatively simple semi-empirical expressions for the distribution function $\varphi(\rho x)$ was proposed by Philibert [68]. After substituting these expressions into the formula (1.23), we obtain the following expression for the correction which takes into account the absorption of radiation in the specimen:

$$f(\chi) = \frac{1}{\left(1+\dfrac{\chi}{\sigma}\right)\left(1+\dfrac{h}{1+h}\dfrac{\chi}{\sigma}\right)}, \qquad (1.24)$$

Here $h = 1.2\dfrac{A}{Z^2}$, $\sigma = \dfrac{4.5\cdot 10^5}{E_0^{1.67} - E_q^{1.67}}$, is the coefficient of attenuation of the electron beam (Lenard coefficient), E_0 is the energy of the electrons, falling on the specimen.

Regardless of the availability of the more accurate but also more complicated relationships for the adsorption correction, the Philibert equation is used widely at present in the calculation of the intensity of characteristic x-ray radiation.

The resultant correction $f(\chi)$ for attenuation is equivalent to the probability F_j used in formula (1.7). Taking into account the generally accepted notations for the correction for attenuation of x-ray radiation in the specimen in the electronic excitation variant, we have retained the notation $f(\chi)$ for the investigated case.

1.4.3.4. Taking into account the effect of selective excitation of target atoms

Probability F_i of ionisation of the atoms during the interaction of the electron beam with the specimen should take into account possible processes of excitation of the atoms by the bremsstrahlung of the electrons and the radiation of other atoms, present in the target (the effect of selective excitation). Taking into account the excitation of the atoms by bremsstrahlung of the electrons, in (1.7) the probability F_i of direct excitation of the atoms is replaced by the total probability of such excitation F_i':

$$F_i' = F_i + F_f = F_i\left(1+\frac{F_f}{F_i}\right) = F_i(1+\Delta), \tag{1.25}$$

Here F_f is the probability of ionisation of the atoms by bremsstrahlung. Under the effect of the electron beam on the specimen, the source of exciting primary irradiation is represented by a thin surface layer. This assumption is explained by the low penetrating capacity of the electrons in comparison with the penetrating capacity of x-ray radiation. For the values of Δ, the following approximate equation was proposed in [69]:

$$\Delta = 0.82\cdot 10^{-5} Z^2 r_q \frac{E_q}{g_q b_q}, \tag{1.26}$$

where $r_q = (S_q-1)/S_q$ and S_q is the adsorption 'jump' (as regards the adsorption 'jump', see section 4 .1).

The values of Δ, calculated from (1.26) for several lines of the x-ray spectrum excited by the electron flux are presented in Table 1.9.

Table 1.9 shows that the contribution of bremsstrahlung to the excitation of atoms for Kα-radiation of the elements with high atomic numbers may be significant.

The expression for the effect of selective excitation of the atoms of the specimen by radiation of other atoms, present in the specimen, is more complicated. Figure 1.10 shows the dependence of the value of the contribution I_f/I_i of the effect of selective excitation to the intensity of x-radiation of iron on the nickel content in the target. Irradiation of the target effectively excites the iron atoms.

Table 1.9. Values of the corrections Δ for the excitation of atoms by bremmsstrahlung for several lines

Line	CrKα	CuKα	RhKα	RhLα	WLα
Δ	0.02	0.05	0.28	0.006	0.03

Figure 1.10 shows that in the investigated case, the contribution of the excitation effect is very significant.

Summing up the results obtained in this section, it may be concluded that the intensity of x-ray radiation of the single-element specimen in excitation of radiation by the electron beam is described (see [70]) by the expression:

$$N_i = \left(\frac{n_q \times b_q}{2 \times 4\pi} \frac{1}{Z \times \ln X} (U_0 \times \ln U - U_0 + 1) \omega_i p_{ij} \right) (1 + \Delta) R_i f(\chi), \quad (1.27)$$

where R_i is the coefficient of backscattering of the electrons, Δ is the correction for the additional excitation by bremsstrahlung, and $f(\chi)$ is the correction for adsorption, given by the expression (1.24).

In (1.27), the intensity is represented by the number of photons in the unit solid angle in calculations for the single incident electron [photons/(electron·s·steradian)].

1.4.3.5. Simulation of the processes of excitation of x-ray radiation by the method of statistical evaluation

Difficulties in the determination of the intensity of x-ray radiation, associated with the ambiguity of the directions of movement of the electrons in the target, taking their backscattering into account, with the determination of some effective depth of formation of radiation, and with calculation of the absorption of the radiation exited by the specimen may be overcome to some extent if we use the method of statistical evaluation (the Monte Carlo method) of the behaviour of the individual electrons in matter. This method was used for the first time by Green [71]. The method was developed further in a number of investigations (see, for example [72–75]).

When using the Monte Carlo method, it is assumed that the specific interactions, accompanying every elementary act of collision and scattering

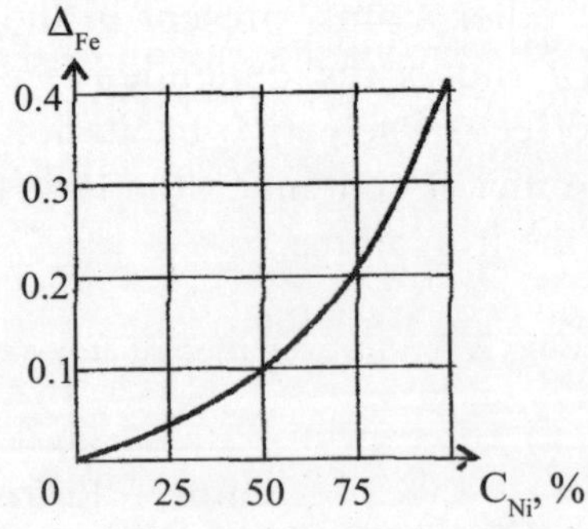

Fig. 1.10. Excitation of iron atoms by radiation of nickel during the interaction of the electron flux with an Fe–Ni target (electron energy E_0 = 25 keV, sampling angle ψ = 40°).

of the electrons, have been studied sufficiently.

The ionisation cross-section is given by the Bethe formula (1.14); the single scattering cross-section is given by the Rutherford formula; the cross-section of interaction with the atom nuclei in braking of the electrons is determined by the Cramer's formula (see chapter 2). Simulation of the trajectory and the energy losses of the particles may be carried out by defining the random numbers γ_c which determine:

1. The energy losses in collision

$\Delta E = Jk \ln \gamma_c$, where J is the ionisation potential;

2. The free path of the electron

$$L = -\frac{\ln(1-\gamma_c)}{N\rho\sigma_s},$$ where σ_s is the scattering cross-section;

3. The angle ϑ of the deflection of the electron in collision

$$\sin^2\left(\frac{\vartheta}{2}\right) = \frac{\gamma_c}{k(1-\gamma_c)+1}, \quad \text{where } k = \frac{8mEr^2}{(h/2\pi)^2}.$$

The summation of a large number of possible trajectories of the electrons and of their interaction with the electrons and atom nuclei makes it possible to determine the macroscopic characteristics of the electron beam penetrating into the specimen and the resultant x-ray radiation.

In the calculations, the trajectory of the electron is treated as a zigzag shaped trajectory, consisting of straight sections whose length is determined by the free path of the electron between two consecutive collisions. The variation of the energy and direction of movement of the electrons in collisions are taken into account. The type of collision, the free path and the possible direction of movement after a collision are determined in accordance with the transverse cross-section of the interactions. Examination of the trajectory of movement of the electron is completed when the electron energy becomes lower than the ionisation energy of the q-shell of the atom or if the electron travels outside the surface of the target, i.e. backscattering takes place.

The statistical models, representing the result of multiple simulation computing experiments, are used usually to determine statistically justified relationships or to verify the accuracy of calculations, carried out using the available approximate analytical expressions.

Chapter 2

Bremsstrahlung

Continuous (braking, bremsstrahlung) x-ray radiation forms during deceleration of charged particles (electrons, protons, ions) in matter since according to the laws of a classic mechanics, a charged particle, moving with acceleration, is a source of electromagnetic radiation.

2.1. Spectral distribution of the intensity of bremsstrahlung

2.1.1. Basic equation for the spectral intensity of bremsstrahlung

The first measurements of the intensity of continuous x-radiation were taken by Moseley and Darwin [76] in 1913 immediately after deriving the relationship which allows the x-radiation to be decomposed into a spectrum using monocrystal structures (the Wulf–Bragg equation). In 1915, Duane and Hunt detected [77] the high-frequency boundary of continuous x-radiation. Studies of continuous x-ray spectra was continued by a number of authors (see, for example [79, 80]). Generalising the collected data and the results of his own investigations, Kulenkampff proposed in 1922 [81] the following equation for the spectral intensity I_ν of the continuous x-radiation of a thick specimen:

$$I_\nu = C\left(Z\left(\nu_0 - \nu\right) + bZ^4\right), \tag{2.1}$$

where Z is the atomic number of a thick target, ν is spectral frequency, ν_0 is the spectral frequency, corresponding to the shortwave boundary of the

---

*After explaining the nature of this radiation, it was possible (see [78]) to determine with sufficient accuracy the ratio of the constants h/e using the equality $eV = h\nu_0$, where the shortwave boundary ν_0 is determined by experiments for the given potential V).

spectrum, C and b are constants.

According to Kulenkampff's data, the second term in the brackets is so small that its effect is appreciable only at frequencies close to ν_0. Later, it was shown that equation (2.1) does not have the second term.

The theory of formation of heterogeneous x-ray radiation, based on the assumption on the deceleration of the electrons in matter, was proposed by Kramers [82] a year later (in 1923). Kramers examined the movement of an electron in the field of the atomic nucleus with the charge Ze. The electron, moving along the hyperbolic trajectory, is accelerated during the Coulomb interaction:

$$a = \frac{Ze^2}{mr^2},$$

where m is the electron mass, r is the distance of the electron from the nucleus.

The electromagnetic energy, emitted by the electron, creates a field with the strength:

$$E = H = \frac{ea}{c^2 R} \sin \varphi,$$

at distance R from the area of interaction of the electron with the nucleus. Here φ is the angle between the acceleration vector a and the radius-vector R.

The intensity of radiation at the observation point at the distance R is determined by the value of the Umov–Pointing vector:

$$I = \frac{c}{4\pi} EH = \frac{e^2 a^2}{4\pi c^3 R^2} \sin^2 \varphi.$$

The total energy, emitted per unit time in all directions, may be determined by integrating this equation with respect to angles φ. Integration is carried out in the polar coordinates (Fig. 2.1).

The area of the elementary ring-shaped section of the sphere is determined by the expression:

$$ds = 2\pi r R d\varphi = 2\pi R^2 \sin \varphi d\varphi.$$

Using this dependence, the following equation is obtained for the total energy emitted by the electron per unit time:

$$\frac{dW}{dt} = \int_0^{\pi} 2\pi R^2 I \sin \varphi \, d\varphi = \frac{2e^2 a^2}{3c^3}. \tag{2.2}$$

The relevant spectral distribution of x-radiation with respect to frequency depends on the acceleration of decelerating electrons.

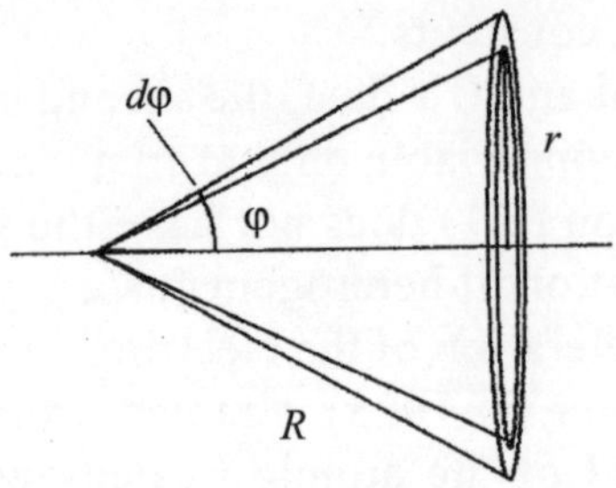

Fig. 2.1. Integration in the polar coordinates with respect to the angles of propagation.

Kramers investigated the components of acceleration a normal and parallel to the vector of the initial velocity of the electron and expanded them into a Fourier series with respect to the frequency of conventional oscillators. As a result of successive transformations, Kramers obtained an equation for the energy of bremsstrahlung $Ed\nu$, formed in the frequency range from ν to $\nu+d\nu$ in s collisions of the electrons with the atoms of the thin-layer of matter:

$$Ed\nu = s\frac{32\pi^2}{3\sqrt{3}}\frac{Z^2e^6}{c^3m^2\upsilon^2}d\nu, \tag{2.3}$$

where υ is the initial velocity of the electron.

In (2.3), the number s of the collisions of the electrons with the atoms of matter for a very thin anode with the area of 1 cm^2 is proportional to the product of the number of atoms M by this area and the number n of electrons incident on the area. Therefore, passing to the intensity of bremsstrahlung (i.e., radiation energy per unit area in 1 s), we obtain:

$$i_\nu d\nu = \frac{16\pi^2}{3\sqrt{3}}\frac{Z^2e^5}{c^3m}\frac{nM}{V}d\nu. \tag{2.4}$$

In deriving this expression, we use the relationship $\frac{1}{2}m\upsilon^2 = eV$, linking the initial velocity of the electrons v with the potential V accelerating the electrons. Equation (2.4) shows that the spectral intensity i_ν of a very thin anode is the same for all frequencies ν and is inversely proportional to the accelerating potential V.

In accordance with equation (2.4), the spectral distribution of bremsstrahlung is extended to any frequency. This result, obtained from the classic theory, was not in agreement with the experimental data available at that time, in particular, with the presence of the shortwave boundary [77] of bremsstrahlung. Therefore, on the basis of the principle of the correspondence of the classic and quantum mechanics representations, Kramers accepted that the appearance of the x-ray photon is caused by the transition of the electron from one hyperbolic trajectory to another. The

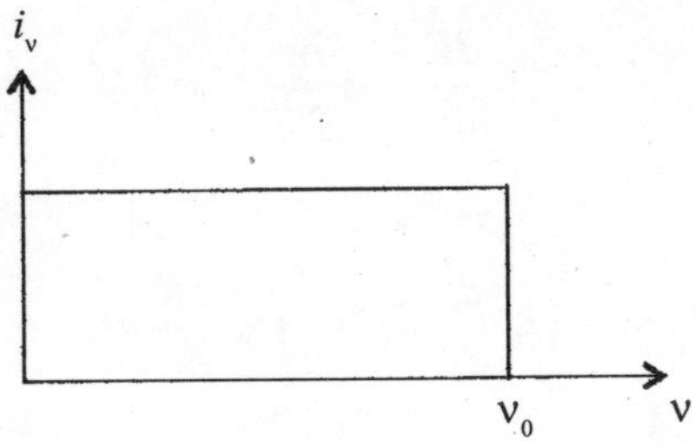

Fig. 2.2. Spectral distribution of the intensity of radiation of a thin target (according to Kramers).

energies of these trajectories are not quantised and, consequently, the energy distribution of the photons is continuous. Since the loss of the electron energy in transition from one trajectory to another cannot exceed the kinetic energy of the electrons, equal to eV, the spectral distribution of bremsstrahlung has a sharp boundary on the side of high frequencies. In other words, the distribution of the intensity of bremsstrahlung with respect to frequencies for the thin anode (the passage of this radiation does not cause any large decrease in electron velocity) should be restricted by the limiting frequency ν_0 (Fig. 2.2).

In order to derive equations for spectral intensity I_ν of bremsstrahlung of a thick anode, it is necessary to integrate equation (2.4) with respect to the depth of penetration of the electrons:

$$I_\nu = \int_0^x i_\nu dx. \tag{2.5}$$

The spectral intensity i_ν of bremsstrahlung of the thin target is described as a function of electron energy. For this purpose, in equation (2.4) the accelerating potential V is replaced by the equivalent value E/e. In addition to this, the number of atoms on the unit surface M is replaced by the number of atoms in the unit volume:

$$\frac{\rho N}{A} \approx \frac{\rho N}{2Z}.$$

Here, we obtain:

$$i_E = \frac{8\pi^2}{3\sqrt{3}} \frac{Ze^6}{c^3 m} \frac{n\rho N}{E}.$$

It is also necessary to take into account that the energy of the electrons E changes with the depth of penetration of electrons into the target. The braking capacity of matter was studied in chapter 1 and is determined by Bethe or Thomson–Widdington laws. If we use the differential form of the Thomson–Widdington law (1.16)

$$\frac{dE}{dx} = -\frac{b\rho}{2E},$$

from (2.5) we obtain:

$$I_E = -\int_{E_0}^{E} i_E \frac{dx}{dE} dE = \int_{E}^{E_0} i_E \frac{2E}{b\rho} dE. \qquad (2.6)$$

Substituting the expression for i_E into (2.6) and after integration, we obtain:

$$I_E = CZ(E_0 - E),$$

where $C = n\frac{16\pi^2}{3\sqrt{3}}\frac{e^6}{c^3 m}\frac{N}{b}$.

Replacing the energy by radiation frequency ($E = h\nu$), we obtain:

$$I_\nu = ChZ(\nu_0 - \nu). \qquad (2.7)$$

Thus, Kramers derived a theoretical expression which coincides with the Kulenkampff's experimental dependence.

The integration of equation (2.7) with respect to frequency makes it possible to determine the integral intensity of bremsstrahlung:

$$I = ChZ\int_0^{\nu_0}(\nu_0 - \nu)d\nu = ChZ\nu_0^2. \qquad (2.8)$$

Equation (2.8) shows that the integral intensity of bremsstrahlung is proportional to the atomic number of the material of the anode and the square of the frequency of the shortwave boundary of the x-ray spectrum (or the square of voltage V in the x-ray tube, since $eV = h\nu_0$).

For the known integral intensity of bremsstrahlung, it is quite easy to determine the coefficient of conversion of the energy of the electron beam to the energy of bremsstrahlung. Actually, the value of C in equation (2.8) is available. Therefore, this coefficient is determined by comparison of the integral intensity with the energy of the electron beam, determined from the equation: $E=nh\nu_0$, where n is the number of the electrons in the beam. Calculations show that only fractions of a percent (in the best case, a percent) of the electron energy are used for the excitation of bremsstrahlung. The main part of electron energy is converted to thermal energy during multiple interactions of the electrons with the electrons of matter.

In a great majority of cases, the expression for the spectral intensity of bremsstrahlung is used in the form of the dependence of the intensity on the wavelength. To obtain the required dependence, it should be taken into account that:

$$I_\lambda = -I_\nu \frac{d\nu}{d\lambda},$$

or taking into account that $\frac{d\nu}{d\lambda} = \frac{d}{d\lambda}\left(\frac{c}{\lambda}\right) = -\frac{c}{\lambda^2}$, according to (2.7) we obtain:

$$I_\lambda = C_1 Z \frac{(\lambda - \lambda_0)}{\lambda^3 \lambda_0}, \tag{2.9}$$

where $C_1 = Chc^2$.

The dependence of spectral intensity on wavelength is shown in Fig. 2.3.

In order to express the spectral intensity of bremsstrahlung through the number of photons N_λ per unit time on unit surface, equation (2.9) should be divided by the photon energy equal to hc/λ. Consequently:

$$N_\lambda = I_\lambda \frac{\lambda}{hc} = C_1 \frac{1}{hc} Z \frac{(\lambda - \lambda_0)}{\lambda_0 \lambda^2}. \tag{2.10}$$

2.1.2. Improvement and modification of Kramers' formula

The previously discussed expressions for the intensity of braking x-ray radiation were derived using the Thomson–Widdington law for the braking capacity of matter. If we use the more accurate Bethe law, given by formula (1.17), we obtain the following equation for the spectral intensity of bremsstrahlung:

$$N_\lambda = \frac{13.65 \cdot 10^{-6}}{4\pi L} Z \frac{(\lambda - \lambda_0)}{\lambda_0 \lambda^2} \left[\frac{\text{photon}}{\text{electron} \cdot \text{steradian} \cdot \text{s} \cdot \text{nm}}\right], \tag{2.11}$$

where $L = \ln\left(\frac{1.166\bar{E}}{J}\right)$, $\bar{E}$ is some mean energy of decelerating electrons, and

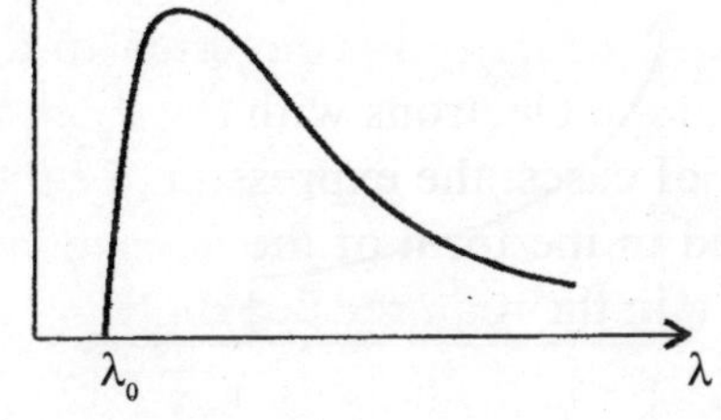

Fig 2.3. Spectral distribution of bremsstrahlung for a thick specimen as a function of wavelength.

$J = 11.5Z$ is the mean excitation energy of the atom.

The mean energy of decelerating electrons depends on the energy E associated with wavelength λ of the spectral distribution. A number of simple equations have been proposed for the mean electron energy, and one of them has the following form:

$$\overline{E} = \frac{E_0 + 2E}{3}.$$

Thus, when using the Bethe law for the braking capacity of matter, the quantity C_1 in the Kramers formula is no longer a constant and depends on the wavelength λ of the spectral intensity of bremsstrahlung. In addition to this, the proportionality between the spectral intensity I_λ and the atomic number Z of the target material, obtained by Kramers, is violated.

The model of the process of deceleration of the electrons, proposed by Kramers, does not take into account the screening effect of the electrons of the atom shell. In addition to this, the classic representation of the braking electrons in the form of conventional oscillators is approximate. Therefore, the fundamental assumption in the Kramers theory that the spectral intensity of bremsstrahlung of a very thin anode does not depend on radiation frequency is also approximate. Sommerfeld [3], taking into account the assumptions of quantum mechanics, carried out calculations of the cross-section of bremsstrahlung of the electrons and studied the scattering of an electron beam, represented by an infinite flat wave in the Coulomb field of the nucleus. The complicated equation, derived by Sommerfeld, indicates that the special intensity of bremsstrahlung of a very thin anode is not constant and increases in a non-linear manner with a decrease of frequency (Fig. 2.4).

Therefore, to take into account possible distortions of the spectral dependence of the bremsstrahlung of a thick anode, a number of authors (see, for example [83–85]) (proposed to use the modified Kramers equation in which the difference multiplier is raised to some power p, close to unity:

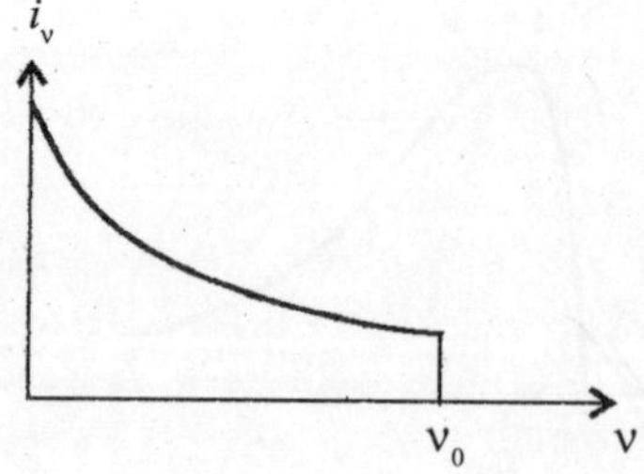

Fig. 2.4. Spectral distribution of braking x-ray radiation of a thin target (according to Zommerfeld).

$$N_\lambda = C_1 Z \frac{(\lambda - \lambda_0)^p}{\lambda^2 \lambda_0}. \tag{2.12}$$

The exponent p depends on the atomic number Z and voltage V in the x-ray tube.

Another variant of the modification of the Kramers equation for thick specimens which takes into account the spectral distribution of bremsstrahlung of a thin target, proposed by Sommerfeld, was put forward by Statham [86] and is based on supplementing the equation (2.11) by the co-multiplier λ^a, where the exponent a has the values of 0.26 for Al, 0.23 for Cr, 0.18 for Cu and 0.15 for Au. To ensure that the scaling coefficient and the spectral intensity in the region of the maximum of spectral distribution remain constant, it is rational to use (see [69]) the correction co-multiplier in the form $(\lambda/(2\lambda_0))^a$.

In addition to making a correction for the dependence of the spectral distribution of the bremsstrahlung of a thin target on frequency, Kramers's equation for the thick anode should be supplemented with corrections for the backscattering of electrons and the absorption of resultant bremsstrahlung in the anode material. It should be mentioned that for the bremsstrahlung the backscattering factor R and the Philibert correction $f(\chi)$, given by the expression (1.23), are almost the same as in the case of characteristic radiation. Therefore, they should be introduced into the investigated dependence (2.12). Consequently, the overall equation for the intensity of bremsstrahlung, represented by the number of photons, emitted per unit time, has the form:

$$N_\lambda = \frac{13.65 \cdot 10^{-6}}{4\pi L} Z \frac{(\lambda - \lambda_0)}{\lambda_0 \lambda^2} R f(\chi) \left(\frac{\lambda}{2\lambda_0} \right)^a \left[\frac{\text{photon}}{\text{electron} \cdot \text{steradian} \cdot \text{s} \cdot \text{nm}} \right]. \tag{2.13}$$

The resultant dependence holds for monoelement matter. If a specimen consists of several elements, instead of Z it is necessary to use some mean value of the atomic number $\bar{Z} = \sum_i C_i Z_i$, and the quantities R and $f(\chi)$ become the function of the composition of matter.

In examination of the spectral distribution of bremsstrahlung of the x-ray tubes, it is necessary to take into account the absorption of emitted radiation in the window of the tube. Therefore, we should write:

$$N_\lambda = \frac{13.65 \cdot 10^{-6}}{4\pi L} Z \frac{(\lambda - \lambda_0)}{\lambda_0 \lambda^2} R f(\chi) \left(\frac{\lambda}{2\lambda_0} \right)^\alpha \exp\{-\mu_{Be} \rho_{Be} d_{Be}\},$$

where μ_{Be}, ρ_{Be} and d_{Be} is the mass coefficient of attenuation, the density and thickness of the beryllium window of the x-ray tube, respectively.

The method of semiconductor detection makes it possible at present to obtain in a very short period of time experimental data on the spectral distribution of radiation of x-ray tubes [87, 88] and use the data instead of the theoretical calculations of spectral intensities.

However, taking into account the variety of possible combinations of the target material and accelerating voltage, it has been proposed to use the available experimental data for deriving semiempirical dependences of the spectral distribution of the parameters affecting this distribution.

For the bremsstrahlung, these semiempirical dependences are based (see, for example [85]) on the modified Kramers law with a simple experimental correction on absorption in the anode:

$$N_\lambda = CZ\frac{(\lambda-\lambda_0)^p}{\lambda_0\lambda^2}\exp\{-Q\mu^q\}. \tag{2.14}$$

In (2.14) it is assumed that the backscattering of the electrons influences the effective depth of formation of bremsstrahlung and is taken into account in the experimental determination of the quantities Q and q. Therefore, it is not necessary to introduce into this equation the backscattering coefficient R and Philibert correction $f(\chi)$.

The parameters p, Q and q in (2.14) determine from the condition of the best agreement of the values calculated from this equation and the experimental values of spectral intensity. These parameters are selected by minimisation of the difference of the intensities:

$$\min\sum_\lambda (N_\lambda^{exp} - N_\lambda) = f(p,Q,q), \tag{2.15}$$

where N_λ^{exp} are the experimental intensities, and N_λ are the intensities, calculated from equation (2.14).

2.2. Spatial distribution of bremsstrahlung

As mentioned in the previous section, the intensity of radiation, emitted by the decelerating electron, is determined by the Umov–Pointing vector:

$$I = \frac{e^2a^2}{4\pi c^3R^2}\sin^2\varphi,$$

where φ is the angle between the director of acceleration of the electron and the vector R, directed towards the examiner. For a very thin target, it may be assumed that the directions of acceleration of the electrons correspond to the vector of the initial velocity of the electrons. Consequently, from the given equation it follows that the intensity radiation is equal to zero in the direction of deceleration of the electron and in the reverse direction ($\varphi = 0°$ or $\varphi = 180°$) , and in the direction, normal to the vector of deceleration of the electrons ($\varphi = 90°$) this intensity is maximum. The spa-

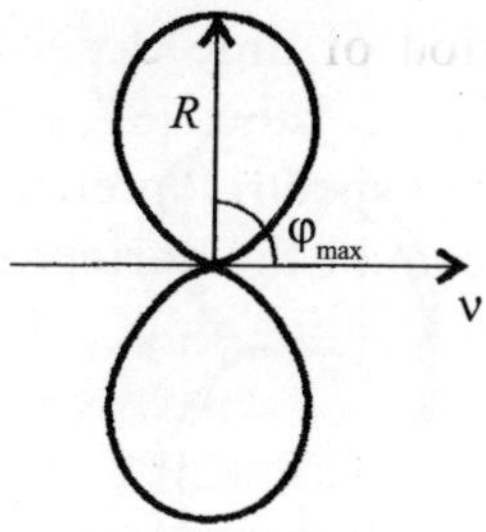

Fig 2.5. Spatial distribution of bremsstrahlung at low velocities of the electron (thin target).

tial distribution of bremsstrahlung for a thin target is shown in Fig. 2.5. This equation holds for non-relativistic velocities of the electrons.

Experimental investigations show [89, 90] that already at a potential of 30 kV, the maximum of the spatial distribution of bremsstrahlung is observed at 55°. Quantum mechanics investigations of the process of deceleration of the electrons, carried out by Sommerfeld in 1931 [91], resulted in the following expression for the angular distribution of the bremsstrahlung of a thin target:

$$I = \text{const}\frac{\sin^2\varphi}{(1-\beta\cos\varphi)^6}, \tag{2.16}$$

where β is the ratio of the velocity of the electron to the velocity of light. A more complicated expression for the special distribution of bremsstrahlung was proposed in [92].

The direction in which the intensity of bremsstrahlung is maximum, is determined by equating the first derivative of (2.16) to zero. Consequently:

$$(\cos\varphi)_{\max} = \frac{1}{4\beta}\left(\sqrt{1+24\beta^2}-1\right). \tag{2.17}$$

The directions, calculated from equation (2.17), and the experimentally determined directions, corresponding to the maximum intensity of bremsstrahlung, are presented in Table 2.1.

The data presented in the table indicate the good agreement between the calculated results and experimental values.

Equation (2.17) shows that at high electron velocities, the maximum intensity of bremsstrahlung is displaced in the direction of movement of the electron and at $\beta \to 1$ the angle $\varphi_{max} \to 0$, i.e., the maximum intensity is directed in the direction of movement of the decelerating electrons. It should be mentioned that at high energies of charged particles (of the order of tens and hundreds of MeV), the spatial distribution of bremsstrahlung is determined by the more complete Bethe–Gaitler theory [4].

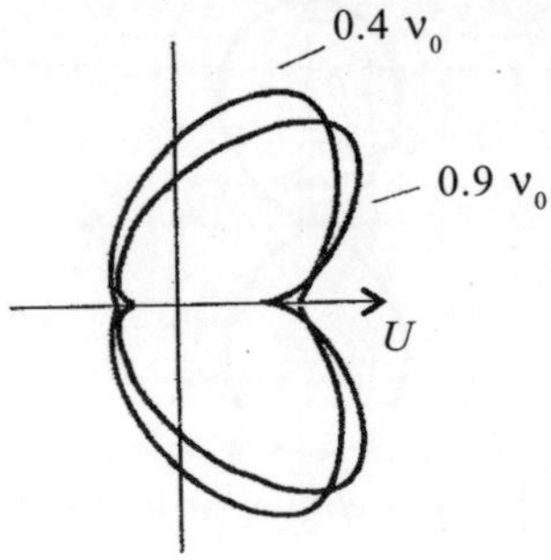

Fig 2.6. Spatial distribution of bremsstrahlung in a thin target at high electron velocities.

Table 2.1. Comparison of theoretical and experimental determined directions of the maximum intensity of bremsstrahlung in the vicinity of its shortwave boundary [2]

Voltage, kV	β	φ^0_{max} (exp)	cos φ_{max} (exp)	cos φ_{max} (theor)
16.4	0.172	63	0.454	0.447
24.0	0.208	59	0.515	0.514
31.0	0.237	55	0.574	0.562

In the emission of a quantum of bremsstrahlung, the electron may lose only part of its initial velocity. Consequently, bremsstrahlung in the frequency range from ν to ν+Δν is determined by the deceleration of the electrons whose velocity changes from β_0 (initial electron velocity) to β_ν (velocity corresponding to the appearance of quanta with frequency ν). Therefore, the maximum intensity for different wavelengths of bremsstrahlung corresponds to different angles φ. This confirms the experimental data, obtained on a thin aluminium foil (V = 34 kV, d = 0.025 μm) and presented in Fig. 2.6. The figure indicates that the experimental spatial distribution of bremsstrahlung better corresponds to the distribution calculated from equation (2.16) for the wavelengths close to the shortwave boundary of bremsstrahlung. At least, at $\nu = 0.9\nu_0$, radiation in the directions φ = 0 and φ = 180° is closer to the zero value expected from this formula. For lower frequencies, the situation is not so clear.

The experimental dependence of the spatial distribution of bremsstrahlung for the frequencies $\nu \rightarrow \nu_0$, determined in [93], is shown in Fig. 2.7. The experiments were carried out on aluminium targets of different thickness with the electron energy of 31 keV. In order to separate the frequencies close to the short-wave boundary and suppress the low-frequency component of bremsstrahlung of the target, radiation was passed through a filter whose thickness is equivalent to 4.5 mm of aluminium. The intensities of

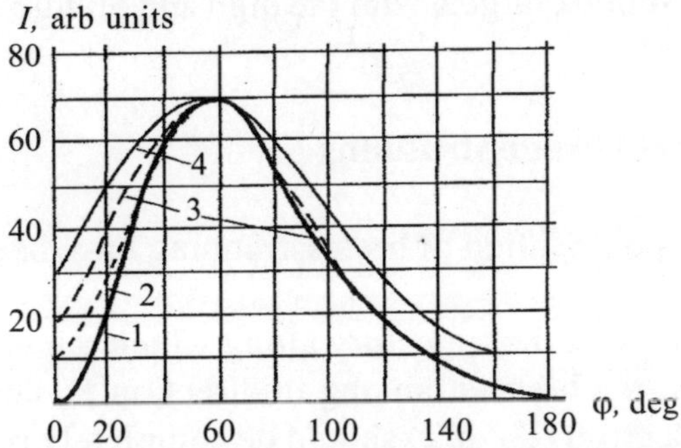

Fig 2.7. Dependence of the spatal distribution of the intensity of bremsstrahlung with the frequency close to the shortwave boundary, on the thickness d of the aluminium target: 1) the theoretical calculations [92] for $d \to 0$; 2–4) experimental data [93] (2 – $d = 10$ μm; 3 – $d = 35$ μm; 4 – $d = 100$ μm).

the distribution maximum for all targets in Fig. 2.7 are added together. A change of the thickness of the target did not change the position of the maximum, indicating that the frequency is close to ν_0.

Thus, for the frequencies close to ν_0, a decrease of the thickness of the target results in the shift of the intensity of radiation in the directions $\varphi = 0°$ and $\varphi = 180°$ in the direction of zero.

In the case of a thick target, the electron beam becomes scattered and the direction of propagation of the beam greatly changes. The degree of scattering increases with increasing of the atomic number of the target. Therefore, the considerable anisotropy in the spatial distribution of bremsstrahlung is detected only in the case of anodes with small atomic numbers (beryllium, aluminium). The spatial distribution of bremsstrahlung, obtained for a thick beryllium anode, is shown in Fig. 2.8.

For anodes with small atomic numbers, experiments showed the same effect as in the case of the thin films: the angles, corresponding to the maximum intensity of bremsstrahlung, differ for different wavelengths.

For high atomic numbers, the anisotropy of bremsstrahlung is not high and equals several percent [86]. Therefore, the spatial distribution of the

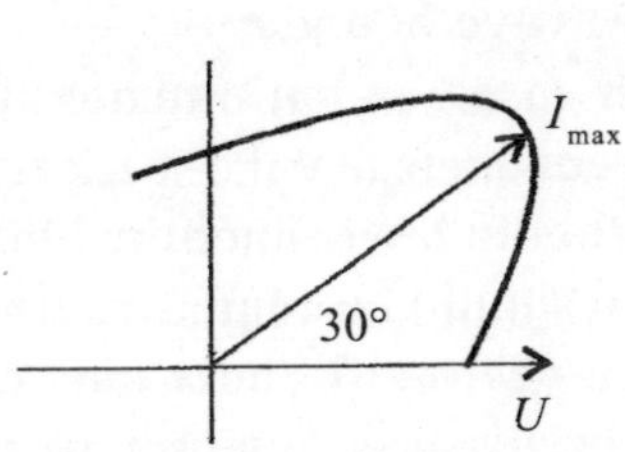

Fig 2.8. The spatial distribution of bremsstrahlung of a thick beryllium anode at a voltage of 140 kV [7].

bremsstrahlung of thick targets with the high and middle atomic numbers is anisotropic.

2.3. Polarisation of bremsstrahlung

The problems of polarisation of bremsstrahlung have been described quite sufficiently in [7].

If it is assumed that the trajectory along which the electrons are decelerated is straight, the bremsstrahlung in direction **R**, normal to the initial velocity of the electron ($\varphi = \pi/2$) should be completely polarised. Actually, the vector of the strength of the electrical field **E** is normal to the radius-vector **R** and is situated in the plane of the vectors **R** and **a** (here **a** is the vector of acceleration of the electron). Consequently, at $\varphi = \pi/2$, the vector **E** is always parallel to the vector of the initial velocity of the electron.

In accordance with the Kramers theory, the electrons decelerate, moving in the field of the atomic nucleus along the hyperbolic trajectories. The shortwave bremsstrahlung forms when the electron passes close to the nucleus and, consequently, the direction of movement of the electron greatly changes. Therefore, the vector of acceleration of the electron in passage through the perihelion of the trajectory is parallel to the vector of the initial velocity of the electron. This case is similar to the previously examined deceleration of the electron along the linear trajectory. Therefore, in the vicinity of the shortwave boundary of the braking spectrum, we should expect a high degree of radiation polarisation.

If the electron passes away from the atomic nucleus and changes its directly only slightly (the frequency of bremsstrahlung is not high), the vector of acceleration of the electron in passage through the perihelion is normal to the vector of the initial velocity of the electron. These cases of interaction of the electron with the nucleus are shown in Fig. 2.9.

Thus, at the frequencies close to the shortwave boundary of the braking spectrum ($\nu \approx \nu_0$), the vectors of the strength of the electrical field and the initial velocity of the emitting electron are parallel ($\mathbf{E} \| \mathbf{v}_0$). At frequencies, far away from the shortwave boundary ($\nu \to 0$), these vectors are perpendicular ($\mathbf{E} \perp \mathbf{v}_0$). In both cases, radiation should be fully polarised. In the case of intermediate frequencies ($\nu_0 > \nu > 0$), partial polarisation should be detected. Of course, it should be remembered that the intensity of polarisation of radiation at $\nu \to 0$ should be considerably lower because in this case the angle φ between the vectors **R** and **a** may change from 0° to 180°.

The polarisation of bremsstrahlung may be measured using as the analyser any substance which efficiently scatters x-rays (for example, graphite). The diagram of this equipment is shown in Fig. 2.10.

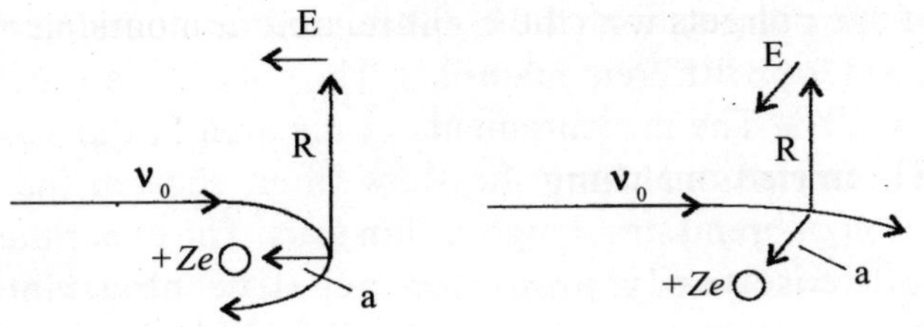

Fig 2.9. Polarisation of bremsstrahlung at frequencies of $\nu \to \nu_0$ and at $\nu \to 0$.

If the radius vector **R**, directed towards the scattering object F, is normal to the vector $\mathbf{v}_0$ of the initial velocity of the electrons, then according to the classic theory the electrons of the scattering solid start to move and the direction of the movement coincides with the direction of the vector **E**. The maximum intensity of scattered radiation is detected in the direction normal to the vector **E** (scattering theory will be examined later). It is assumed that the plane, passing through the scattering objects and being normal to the radius vector **R**, contains two detectors, recording scattered radiation. The direction to one of these detectors is normal to the vector of the initial velocity of the electrons (for measuring intensity $I_\perp$) and to the other detector it is parallel to the vector of their initial velocity (for measuring intensity $I_\parallel$). Consequently, the ratio of the intensities of scattered radiation, recorded by these detectors, characterises the degree of polarisation of bremsstrahlung. Actually, the shortwave completely polarised bremsstrahlung ($\mathbf{E}\|\mathbf{v}_0$) after scattering on the analyser can be recorded only by the detector $D_{E\perp v0}$ which measures $I_\parallel$. The degree of polarisation is characterised by the polarisation coefficient P:

$$P = \frac{I_\perp - I_\parallel}{I_\perp + I_\parallel}. \tag{2.18}$$

The previous considerations regarding the polarisation of bremsstrahlung have been confirmed by the calculations of the polarisation coefficient, carried out in [94], using the Sommerfeld quantum mechanics theory. The resultant dependence of the polarisation coefficient P on angle φ for different frequencies is shown in Fig. 2.11.

Figure 2.11 shows that at frequencies $\nu \to \nu_0$ the polarisation coefficient is very close to unity at angles φ close to 90°. For low frequencies ($\nu \to 0$), polarisation is negative and approaches –0.6 at $\varphi \to 90°$. Polarisation depends on voltage V, accelerating the electrons, and on the atomic number Z of the target material. In the vicinity of the shortwave boundary the degree of polarisation increases with increase of the ratio V/Z^2.

The experimental investigations, carried out on very thin aluminium foils, have confirmed the fact of polarisation of bremsstrahlung. The authors of

[95, 96] measured the polarisation of the entire radiation passing through the thin foil produced from different materials. The polarisation coefficient was approximately 50%. The measurements in the area of the shortwave boundary [97, 98], carried out using the Ross filter, showed the almost complete polarisation of bremsstrahlung for thin foils. The experiments did not confirm the theoretically predicted negative polarisation of bremsstrahlung in the low frequency range (in Fig. 2.11, the dependence for $\nu \to 0$) using the scheme shown in Fig. 2.10 (in the direction of the detector $D_{E\perp v0}$).

For thick targets, the radiation is polarised only in the vicinity of the shortwave boundary because the formation of lower frequencies is associated with the deceleration of the electrons which have already lost part of their energy in collisions and changed their initial direction. The dependence of polarisation in the range of the shortwave boundary on the energy of the decelerating electrons has been examined in, for example, [99, 100].

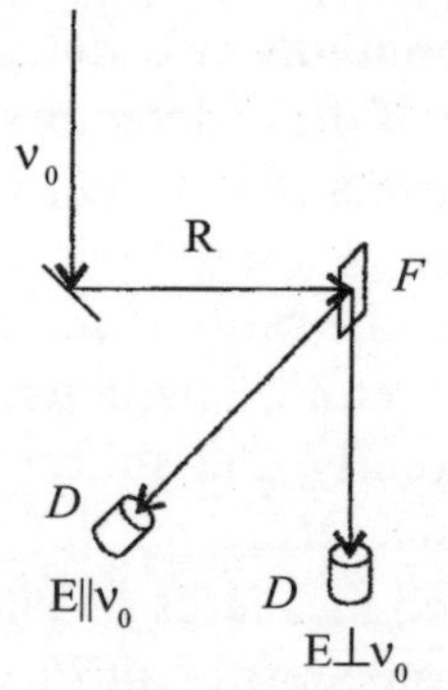

Fig 2.10. The diagram of equipment for recording polarised bremsstrahlung.

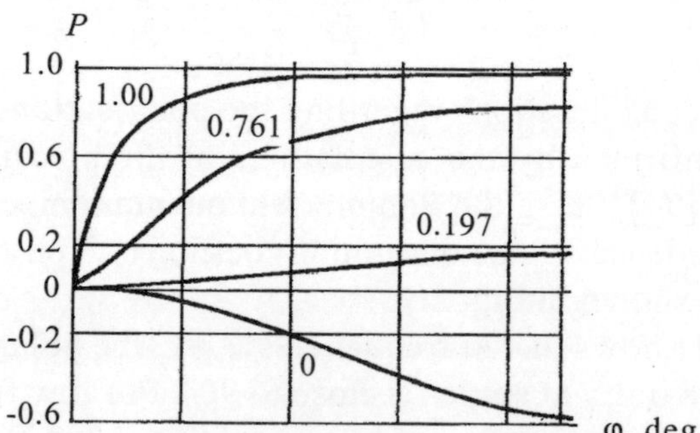

Fig 2.11. Dependence of the polarisation coefficient on the angle φ at $V/Z^2 = 1.776$ at the values of ν/ν_0, shown in the figure.

Chapter 3

Sources of x-ray radiation

The information presented in the previous chapters indicates that the primary x-radiation forms either during ionisation of the internal shells of atoms (characteristic radiation) or during changes in the velocity of charged particles as a result of interaction of the particles with the electromagnetic field (bremsstrahlung). In this chapter, several artificial and natural sources of primary x-radiation are discussed.

3.1. X-radiation excited by the electron beam. X-ray tubes

The electron beam is a tool for studying microvolumes of matter (x-ray microprobe, electron microscope), for examining the structure of materials by x-ray diffraction and, finally, for producing radiation fluxes of x-ray tubes with the controllable spectral composition for use in defectoscopy, medicine, determination of the elemental and structural composition of materials, etc.

The currently available x-ray tubes have the form of a high-power electron lamp (diode). The electron flux, emitted by the cathode (the heating filament is made of a refractory metal), is accelerated by the potential difference applied to the electrodes and generates x-radiation during deceleration on the anode. The voltage between the electrodes amounts to tens (in a small number of cases hundreds) of kilovolts with the current ranging from hundredths of a milliampere to hundreds of milliamperes. Depending on the application, the design of the x-ray tubes differs. The electron beam may be focused into a small area or deconcentrated over the area; high voltage may be applied to the cathode (negative polarity) or the anode of the tube (positive polarity); the window of the x-ray tube, produced from the material with a low absorptivity, which may be situated at the end or on the side surface of the tube; the anode may be thick or have the form of a thin layer of material, deposited on the window of the tube (the tubes with a transmitting anode).

The anode material determines the spectral composition of x-ray radiation and contains the continuous and characteristic components. The following equation may be written for the spectral intensity of the radiation emitted by the x-ray tube:

$$I(\lambda)=\left(I_\lambda+\sum_i \delta(\lambda-\lambda_i)I_i\right)\exp\{-\mu_{Be}(\lambda)\rho_{Be}d_{Be}\}, \tag{3.1}$$

where $\delta(\lambda-\lambda_i)$ is the function having the value of 1 at $\lambda = \lambda_i$, and 0 at all other values of λ. The exponent takes into account the radiation emitted from the tube after its beryllium window. In this case, $\mu_{Be}(\lambda)$ is the mass coefficient of absorption of radiation with the wavelength λ in beryllium; ρ_{Be} is the density of the material; d_{Be} is the thickness of the beryllium window.

The spectral composition of radiation of the x-ray tubes can be greatly varied using filters installed behind the window of the tube. The filters absorb the radiation emitted by the tube to different degrees for different wavelengths so that by the efficient selection of the filter material it is possible to suppress the long-wave or short-wave component of this radiation.

3.1.1. The ratio of the intensities of the characteristic and continuous components of radiation of x-ray tubes

In the theoretical modelling of the spectral intensity of the tubes using the dependence (3.1) it is necessary to know the ratio of the continuous I_λ and characteristic I_i components of polychromatic x-ray radiation. This ratio is determined by definition of the numerical values of the theoretically determined coefficients in the expressions (1.27) and (2.13) for the intensity of these components.

The use of the theoretical relationship between these components assumes that the relationship is in agreement with the experimental data.

In the semiempirical variant of the determination of the ratio of the characteristic and continuous components of primary x-ray radiation [101, 102] it is assumed that the spectral composition of the continuous component is determined by the Kramers equation. Measurements are taken of the intensity of x-ray fluorescence of some element in the presence and absence, in the composition of primary radiation, of the characteristic component which effectively excites the fluorescent element. The appearance of this characteristic component may be determined either by the increase of voltage in the x-ray tube to the value exceeding the excitation potential of the characteristic lines of the anode material [101] or by the replacement of the anode material which does not excite the atoms of the fluorescent element by its characteristic radiation by the anode material capable of exciting these atoms by its characteristic radiation [102]. The resultant differences in the intensity of fluorescence in the presence and absence of the characteristic component, exciting fluorescence, make it possible to calculate the required ratio of the continuous and characteristic components of primary radiation.

In the second variant (see, for example [103, 104]) the ratio of the intensity of the characteristic line of the primary spectrum to the spectral intensity of bremsstrahlung below this line is determined by the simultaneous measurement of the characteristic and continuous components of primary radiation using a semiconductor detector.

The dependence of the ratio of the intensities of the characteristic and continuous components of primary x-radiation on the atomic number of the anode material and voltage in the x-ray tube is determined using the equations (1.27) and (2.13). From (2.13), ignoring the correction $\left(\frac{\lambda}{2\lambda_0}\right)^a$, we obtain

$$N_\lambda = AZ\left(\frac{\lambda/\lambda_0 - 1}{\lambda^2}\right)Rf(\chi),$$

where A is a constant. Since we are interested in the spectral intensity of bremsstrahlung corresponding to the line i of the characteristic spectrum, in the equation it should be assumed that $\lambda = \lambda_i$. Since $\lambda/\lambda_0 = \lambda_i/\lambda_0$, or assuming in the first approximation that $\lambda_i \approx \lambda_q$ (here λ_q is the wavelength, corresponding to the binding energy of the electrons of the q-shell), it is obtained that $\lambda/\lambda_0 \approx \lambda_q/\lambda_0 = V_0/V_q = U_0$. This means that:

$$N_\lambda \approx AZ\frac{U_0 - 1}{\lambda_i^2}Rf(\chi).$$

Examining the ratio of the intensities N_i/N_λ, where the intensity of characteristic radiation N_i is given by the expression (1.27), without the correction for the additional excitation of the lines by bremsstrahlung, we obtain:

$$N_i/N_\lambda = \frac{I_i}{I_\lambda} \approx B\frac{1}{Z^2}\left(\frac{U_0 \ln U_0}{U_0 - 1} - 1\right)\omega_i\lambda_i^2, \tag{3.2}$$

here B is a constant. According to equation (1.12), proposed by Wentsel, the fluorescence yield ω_i increases approximately proportionately to Z^4. On the other hand, in accordance with the Moseley equation (see equation (1.3)), the quantity λ_i^2 decreases in the first approximation in proportion with Z^4. Therefore, for the qualitative evaluation of the investigated ratio

Table 3.1. The ratio of the intensities of characteristic radiation and bremsstrahlung for several melts and different voltages of the x-ray tube

Voltage [kV]	Anode, line			
	CrKα	RhKα	$RhL\alpha_{1.2}$	$WL\alpha_{1.2}$
30	6.70	0.07	4.47	0.58
40	7.55	0.13	4.98	0.72
60	9.62	0.23	5.53	0.93

the value $\omega_i \lambda_i^2$ may be regarded as independent of Z. Equation (3.2) shows that:

1. The contrast of the characteristic lines rapidly decreases with the increase of the atomic number of the anode material.
2. The contrast increases with the increase of x-ray tube voltage.

Table 3.1 gives the experimental values of the relative intensities N_i/N_λ for chromium, rhodium and tungsten anodes.

The data presented in the table are in agreement with the theoretically predicted dependence of the relative intensity on the atomic number and the voltage in the x-ray tubes.

3.1.2. Special features of the spectral distribution of radiation of the x-ray tubes with an earthed cathode

The high potential of the anode of the x-ray tubes returns the backscattered electrons [6, 105] preventing their destructive effect on the tube window. The anode may come closer to the tube window and the thickness of the window may decrease. This results in a large increase of the intensity of emitted long-wave radiation. Therefore, the application of x-ray tubes with a high anode potential is a widely observed action.

As mentioned in the section 1.4.3.2, the ratio of the number of the backscattered electrons to the number of the electrons of the cathode beam (the backscattering coefficient η) may reach 50%. Therefore, the return of the backscattered electrons to the anode surface greatly changes the intensity of bremsstrahlung and characteristic radiation [106, 107], represented by the relationships (1.27) and (2.13). We examine this problem in greater detail.

Let the number of these electrons in the energy range from E to $E+\Delta E$ be:

$$n_1 = \left(\frac{dn}{dE}\right)_1 = \frac{n_0}{E_0}\Phi(E, E_0, Z), \tag{3.3}$$

where n_0 and E_0 are respectively the number and energy of the electrons of the cathode beam; $\Phi(E, E_0, Z)$ is the distribution function of the backscattered electrons with respect to energy. In [61], the following equation was proposed for this function:

$$\Phi(E, E_0, Z) = a\left(0.27\left(\tfrac{E}{E_0}\right)^{0.02Z} + \left(1.1 + \tfrac{5}{Z}\right)\left(\left(\tfrac{E}{E_0}\right)^{0.1Z} - 1.1\left(\tfrac{E}{E_0}\right)^{0.4Z}\right)\right),$$

where

$$a = Z(-0.00391\ln(Z+1) + 0.00721(\ln(Z+1))^2 - \\ - 0.001067(\ln(Z+1))^3)F(E_0) \tag{3.4}$$

and $F(E_0) = \left(\dfrac{E}{30}\right)^{-0.25+0.0787Z^{0.3}}$.

After the secondary interaction with a positively charged anode, part of the electrons of the first generation diffuse again through the surface, generating a second-generation of the backscattered electrons whose energy distribution can be determined by integration of the equation (3.3) with respect to the energy of the electrons of the first generation in the range from E to E_0:

$$\left(\frac{dn}{dE}\right)_2 = \int_E^{E_0} \frac{n_1}{E_1}\Phi(E, E_1, Z)\,dE_1, \tag{3.5}$$

where n_1 is the number of the backscattered electrons of the first generation with the energy E_1 in the range from E_1 to E_1+dE_1, given by the ratio (3.3)

The energy distribution of all backscattered electrons above the positively charged surface is determined from the equation:

$$\frac{dn}{dE} = \sum_k \left(\frac{dn}{dE}\right)_k,$$

where k is the number of 'rediffusions'.

Figure 3.1 shows the energy distributions of the backscattered electrons of the first and second generations, calculated using equations (3.3) and (3.5) (see [106]).

Figure 3.1 shows that the contribution of the backscattered electrons of the second generation to the total distribution is significant. As expected, the maximum of this distribution is shifted in the direction of low energies in comparison with the maximum of the electrons of the first generation.

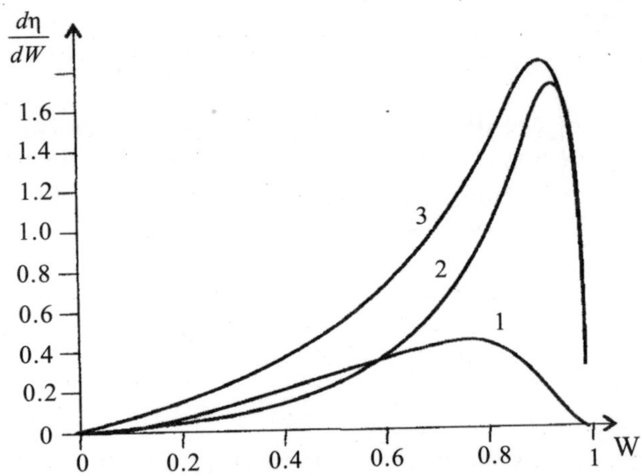

Fig. 3.1. The energy distribution of the backscattered electrons of the first (2) and second (1) generations for the tungsten anode, (3) – the overall distribution. In the graph $W = E/E_0$, $\eta = n/n_0$.

The spectral intensity of bremsstrahlung $N_1(\lambda)$, determined by the effect of the backscattered electrons, is determined as the integral:

$$N_1(\lambda) = \int_{E_\lambda}^{E_0} N(\lambda, E)\left(\frac{dn}{dE}\right) dE, \tag{3.6}$$

where E_λ is the energy, corresponding to the wavelength of bremsstrahlung, $N(\lambda, E)$ is the spectral intensity of bremsstrahlung (given by the equation (2.13)) at the energy E of the electrons exciting this radiation.

The spectral distribution of the intensity of bremsstrahlung of a tungsten x-ray tube with the earthed cathode at a voltage of 40 kV is indicated by the dependence 4 in Fig. 3.2. The contribution of the electrons of the cathode beam is determined by the curve 3. The contributions of the backscattered electrons of the first and second generation, calculated using equation (3.6), are indicated by the relationships 2 and 1, respectively.

If it is assumed that $N(\lambda, E)$ determines the intensity of characteristic x-ray radiation (given by the equation (1.27)) at the energy E of the exciting electrons, and if the lower limit of integration is assumed to be equal to the binding energy E_q on the q-shell, the equation (3.6) may be used to calculate the intensity of the characteristic radiation excited by the backscattered electrons.

The ratio of the intensities of the lines of the characteristic spectrum of the chromium, rhodium and tungsten anodes with the earthed cathode and the anode, shown in Table 3.2.

Table 3.2 shows that for the anodes with the high atomic numbers, the

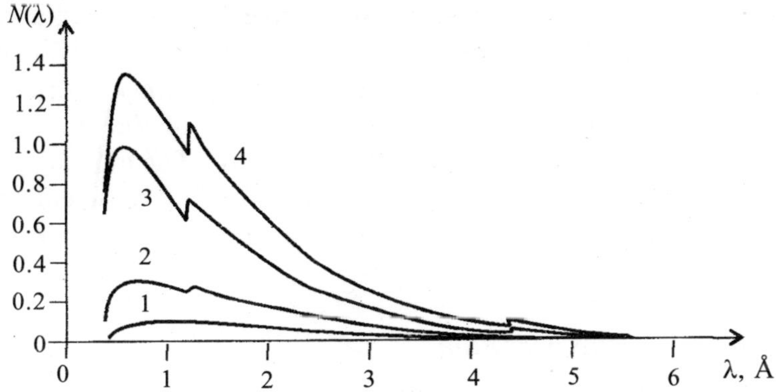

Fig. 3.2. Contributions of the electrons of the cathode beam (dependence 3) and of the backscattered electrons of the first (relationship 2) and second (the relationship 1) generations in the formation of the braking spectrum of the tungsten anode. Only the most significant L_3- and M_5-radiation jumps are presented. The dependence 4 is the overall spectral distribution for the earthed cathode.

Table 3.2. The ratio of the intensities of the α-lines of chromium, rhodium and tungsten anodes with the earthed cathode (N_i) and the earthed anode (N_i') of the x-ray tube

Radiation	CrKα		RhKα		RhLα		WLα	
λ [nm]	0.229		0.062		0.460		0.148	
Voltage [kV]	40	60	40	60	40	60	40	60
N_i/N'_i	1.16	1.18	1.15	1.22	1.40	1.45	1.41	1.48

transition to the x-ray tubes with the earthed cathode increases the intensity of the long-wave characteristic radiation by a factor of almost 1.5. The contrast of the characteristic lines changes only slightly because the spectral intensity of bremsstrahlung of the x-ray tube in the area of these lines increases when the cathode is earthed by approximately the same amount.

3.2. Excitation of x-radiation by the ion beam

X-radiation may be excited by an ion flux, i.e., charged particles whose mass is considerably larger than the electron mass. This is accompanied by the formation of both the characteristic and bremsstrahlung.

Powerful fluxes of the charged particles with high energies are usually produced in direct action accelerators with the single passage of particles through a large potential difference [108].

The main theoretical principles of deceleration of the ions in matter were formulated by Bohr [109]. The processes of formation of x-radiation in the ion excitation of atoms have been examined in sufficient detail in [110].

In passage through the matter, the ions collide mainly with the atomic electrons and lose their energy for the ionisation of the atoms, without deviating from their trajectory (inelastic collisions). The trajectory changes only in the case of elastic collisions with atom nuclei. These collisions are relatively rare and, consequently, we shall consider electronic deceleration of the ions.

3.2.1. Characteristic radiation

The ion beams interact on the whole with the atom of matter and not with individual shell electrons. The energy, lost by the ion, is carried away by one of the electrons of the atom.

The intensity of characteristic x-radiation in excitation with atoms is determined, as in the case of excitation of the atoms by the electrons, by the dependence:

$$I_i = \frac{\rho N}{A} \int_x Q_q(E) \frac{dE}{dx} dx,$$

where $\frac{\rho N}{A}$ is the number of atoms in the unit volume, $Q_q(E)$ is the ionisation cross-section of the q-shell of the atom, $\frac{dE}{dx}$ is the braking capacity of the target (electronic deceleration).

For ion bombardment, the ionisation cross-section and the braking capacity of the target differ from the previously examined case of interaction of the electron beam with the matter.

The ionisation cross-section Q_q of some atomic level q in the first approximation is proportional to the square of the atomic number Z_δ of the incident particles, inversely proportional to the fourth power of the atomic number Z of the bombarding matter and depends on whether the energy of the particles E is higher than the binding energy E_q of the electrons on the q-level (see, for example [111]):

$$Q_q \propto \frac{Z_\delta^2}{Z^4} f\left(\frac{E}{E_q}\right), \tag{3.7}$$

where the function $f\left(\frac{E}{E_q}\right)$ is presented in the form of tables (see, for example, [112]).

Detailed examination of the ionisation cross-section of the internal shells

[113] assumes consideration of the energy loss and Coulomb deflection of the incident ion, and also the excited stationary state and the relativistic nature of the internal shells of the system. It was shown in [114] that the results of calculations of the ionisation cross-section of the K-shell for protons are in good agreement with the results of experimental investigations collected in the databases used in practice for these values.

Several expressions have been proposed for electronic deceleration *dE/dx*. For low ion energies, it is usually recommended to use the relationship proposed in [115], and for higher energies the Bethe–Bloch equation, presented in [116].

The relationship (3.7) shows that to increase the intensity of the x-radiation, it is necessary to use the ions with high atomic numbers and high energies. However, in these cases it must be taken into account that bombardment with heavy ions results in rapid heating and failure of the irradiated specimen. Therefore, x-radiation is excited in most cases by the ions with low values of Z_δ (for example, protons).

In addition to this, equation (3.7) shows that ion excitation is most efficient for low atomic numbers *Z*. Table 3.3 gives the experimental data [117], characterising the ionisation capacity of several ions (at the ion energy of 75 keV), for the copper and carbon atoms.

Table 3.3 shows that the ionisation capacity of the ions for the carbon atoms is five orders of magnitude higher than that for the copper atoms. The increase of the ionisation capacity with the increase of the atomic number of the ions at the energy of 75 keV is not so significant.

Figure 3.3 shows the dependence of the ionisation capacity of the α-particles of different energies on the number *Z* in the excitation of K series of x-ray radiation.

It may be seen that, for example, at $Z = 20$, the yield of x-ray radiation increases approximately 100 times when the energy of α-particles is increased from 2 to 6 MeV.

For protons, the energy of the order of 1–5 MeV is optimum in excitation of the atoms.

Table 3.3. The number of photons/ions in the excitation of x-ray characteristic radiation of carbon and copper by the ion flux with different atomic numbers

Excited element	Exciting ions			
	H	He	N	Ar
C	$1.5 \cdot 10^{-4}$	$1.4 \cdot 10^{-4}$	$2.8 \cdot 10^{-4}$	$9.8 \cdot 10^{-4}$
Cu	$1.3 \cdot 10^{-9}$	$2.1 \cdot 10^{-9}$	$3.0 \cdot 10^{-9}$	$5.6 \cdot 10^{-9}$

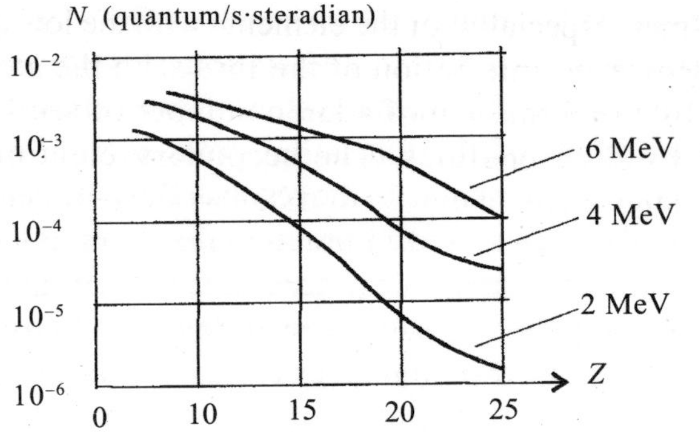

Fig. 3.3. Effect of the energy of α-particles on the yield of characteristic x-radiation for different elements.

It should be mentioned that in ionisation of the atoms by the ion beam with high atomic numbers (an exception are protons and ions of the helium atoms), the recorded spectrum contains, in addition to the lines of the element of the target material, also the characteristic lines of the incident ions.

Another special feature of the spectra, excited by the ions, is the shift of the lines in the direction of short wavelengths. This is caused by the fact that the appearance of vacancies in the internal shells is accompanied by the formation of vacancies in the outer shells of the atom and this increases the binding energy of the internal electrons in the atom.

Ion excitation is usually used for the examination of the surfaces and subsurface layers of matter because the depth of penetration of the ions is not large.

Since the contrast of the characteristic lines is determined by the size of the background, it is important to examine the problem of formation of background bremsstrahlung in ion bombardment.

3.2.2. Bremsstrahlung of ions

The bremsstrahlung of the ions forms as a result of their Coulomb interaction with the atomic nucleus. The intensity of this radiation is determined by the quantum mechanics theory of dipole radiation [118], according to which the ions generate radiation like electrons in interaction with the atom nuclei. However, for the ions, the probability of emission of bremsstrahlung is approximately a factor of $(M/m_e)^2$ smaller in comparison with the electrons (M is the mass of the ion; m_e is the electron mass). The mass of the lightest ion (proton) is 1836 times greater than the electron mass. Therefore, it should be expected that the x-ray background for the characteristic lines of primary deviation in ion excitation is very small, and the contrast of the characteristic lines, especially for the elements with the low atomic numbers, is very high. However, interaction of the ions with the atoms of matter is accompanied by the formation of a large number of secondary electrons, characterised by high energies. The secondary electrons in Coulomb interaction with the nuclei of the atoms of the target generate bremsstrahlung (chapter 2). In addition to this, the incident particle may ionise the internal shells of the atoms of the target, transferring the knocked-out electrons into the region of continuous state. These electrons, returning to their initial condition, emit photons with a continuous energy spectrum. This radiation is referred to as atomic bremsstrahlung. Theoretical expressions for the cross-section of these processes of formation of bremsstrahlung have been published in [119], for example. The relationship between these main components of bremsstrahlung, as a function of the atomic number of the target and proton energy, has also been investigated in [119].

The investigated bremsstrahlung is many times stronger than the bremsstrahlung of the spectrum of the ions (see, for example [120, 121]). Figure 3.4 shows the dependence of the cross-section Q of the braking interaction of the protons and the secondary electrons (related to the binding energy E_q of the electrons in the ionised atoms) on the energy of this bond, which is proportional to the atomic number Z of the element [122].

Figure 3.4 shows that in the range of characteristic lines of the elements with the low atomic numbers, the bremsstrahlung of the secondary electrons is many times higher than the bremsstrahlung of the protons which becomes controlling only at the energies of $E_x > 10$ keV, where the ionising capacity of the ions is not so effective. This confirms the dependence of the quantity, proportional to the limit of detection, on the atomic number Z of the excitation element, published in [122] and shown in Fig. 3.5.

Figure 3.5 shows that there is the optimum range of the excited elements where the value C, proportional to the limit of detection, is minimum. At low atomic numbers, the background determined by the bremsstrahlung of the secondary electrons is very significant, and at higher values of Z the rapid decrease of the ionisation cross-section with the increase of Z of the excitation element is important.

This figure also shows that the detection limit for the optimum range is less than 10^{-4}–10^{-5}%. This means that in the ion method of excitation of the atoms, we can determine the relatively low content of the elements in different matrices.

3.3. Radioactive sources of x-ray radiation

The main radioactive sources have been described in detail in [123–125].

The radioactive isotopes are produced by the irradiation of targets with

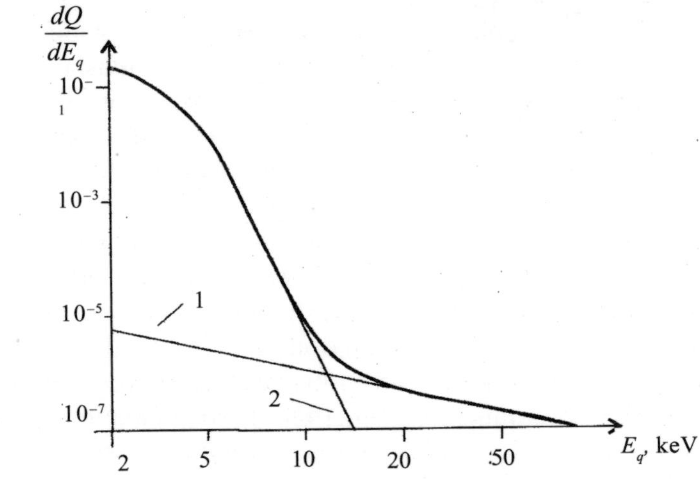

Fig. 3.4. Excitation of bremsstrahlung by protons with the energy of 2 MeV (dependence 1) and secondary electrons (dependence 2) as a function of the binding energy of the electrons in the atom. Observation under the angle of 90° in relation to the direction of incidence of the proton beam.

charged particles in a cyclotron or in nuclear reactors.

The radioactive isotopes emit characteristic and/or bremsstrahlung. Three variants of the formation of x-ray radiation are possible here:

– internal conversion or the capture of the electron of the K-shell by the atomic nucleus, with subsequent emission of characteristic K-radiation;

– interaction of β-radiation (i.e., electrons, generated in the process of nuclear transformations) with the atoms of matter leading to the formation of bremsstrahlung and characteristic x-ray radiation;

– the x-ray radiation, accompanying α-breakdown.

Each variant will be examined separately.

3.3.1. Capture of the electron from K-shell by the atomic nucleus. Internal conversion

This type of capture results in the formation of a vacancy on the K-shell, and the charge of the atom nucleus decreases by unity. This is followed by the emission of characteristic x-ray radiation for the element with the atomic number smaller by unity than the atomic number of the radioactive isotope.

The x-ray radiation of these isotopes does not contain the continuous component, and the specific intensity of this radiation is usually high. These isotopes are usually typical by a short half-time period. Table 3.4 gives the parameters of several isotopes of this type [124].

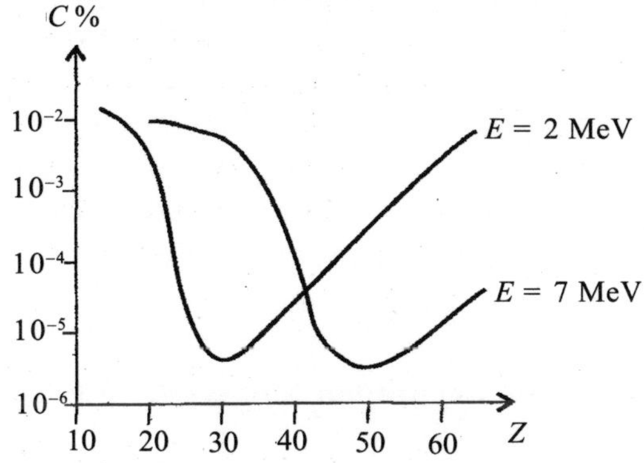

Fig. 3.5. Calculated dependence of the content C of the element on its atomic number Z. The interfering matrix – carbon. The content C is characterised by the value at which the intensity of the K-radiation of the element, excited by the protons, is equal to the intensity of bremsstrahlung.

Table 3 .4. Characteristics of several isotopes, having the K-capture property

Isotope	Halftime	Radiation energy (keV)	Maximum specific activity
^{55}Fe	2.6 years	5.9 (Mn K)	1 Ci*/g
^{57}Co	270 days	6.4 (Fe K) + 122	0.5 Ci/ml
^{93}Mo	2600 years	16.6 (Nb K)	–
^{109}Cd	1.3 years	22.6 (Ag K)+ 88 (4%)	20 μCi/ml
^{181}W	140 days	58 (Ta K) 8 (Ta L)	20 μCi/ml

*Ci (curie) is the unit of isotope activity (number of decays per second), 1 Ci = $3.7 \cdot 10^{10}$ s^{-1}

Table 3.4 shows that the radiation of the investigated ions may also contain a weak short-wave component determined by nuclear processes.

3.3.2. Bremsstrahlung and characteristic radiation of sources, associated with nuclear β-radiation

The isotopes of this type are sources of bremsstrahlung formed during the deceleration of high-energy electrons in the isotope and/or in a filler with which the isotope is mixed. The ionisation of the atoms of the filler results in the characteristic radiation of the source. This means that the spectral composition of the investigated sources contains, in addition to the inevitable bremsstrahlung, the characteristic component.

The source of β-radiation may be simply mixed with a filler (the homogeneous source) and placed in a vessel preventing the exit of the electrons. The isotope may be placed in a capsule used as the target (the source with the passage target). Finally, the source and the target can be separated in space (the source with the reflection target). Table 3.5 shows the parameters of several radioactive sources, based on the application of β-radiation.

3.3.3. X-ray radiation, accompanying α-breakdown

The process of α-breakdown is usually accompanied by the emission of x-rays or γ radiation. In addition to this, the interaction of α-particles with high energies with the atoms leads to the formation (section 3.2) of characteristic x-ray radiation. The highest efficiency is shown by the particles with the energy of the order of 2–6 MeV. The x-radiation of the ^{238}Pu and ^{241}Am isotopes with the energy of α-particles (~5.5 MeV) is presented in Table 3.6.

3.4. Synchrotron – a source of x-ray radiation

The synchrotron is a cyclic accelerator of electrons with the time-dependent controlling field and with a constant frequency of the accelerating electrical field. The synchrotron is a source of magnetically braking (synchrotron) radiation of electrons which travel at relativistic velocities in a homogeneous

Table 3.5. Characteristics of several sources of x-ray radiation with β-breakdown

Source	Halftime	Radiation energy (keV)	Maximum specific activity
^{3}H/Ti	12.3 years	Braking radiation to 18 keV 4.5 (Ti K)	2.6 Ci/cm^2
^{35}S/Ba	85 days	Brading radiation to 167 keV; 32.2 (Ba K)	8 Ci/g
^{90}Sr/Pb	28 years	Braking radiation to 2200 keV; 74 (Pb K)	12 Ci/g
^{147}P/different Targets	2.7 years	Braking radiation to 225 keV; 39 (Pm K); characteristic radiation of targets	10 Ci/g
^{204}Tl	3.4 years	Braking radiation to 760 keV	60 μCi/g

Table 3.6. The x-ray radiation of the isotopes of plutonium and americium formed during their α-breakdown

Isotope	Halftime	Radiation energy [keV]	Maximum specific activity
^{238}Pu	86 years	(13-17) L radiation U 43.5	100 μCi/g
^{241}Am	458 years	(14-18) L radiation Np 20.9; 26 (3 %); 60 (40%)	2 Ci/g

magnetic field. At the relativistic electron velocities, the irradiation of the electrons in the region of the highest harmonics has a continuous spectrum in the range from the radio-frequency range to the x-ray range.

The properties of synchrotron radiation and some of its applications have been examined in, for example [126, 127].

In the synchrotron, the beam of electrons with the energy ε_{el} (of the order of 5–10 GeV) moves in a strong magnetic field (approximately 10^6A/m) around a circular orbit with the radius L (Fig. 3.6).

The acceleration of the electrons under the effect of the magnetic field results in the emission of electromagnetic radiation by the electrons. The intensity of this emission I at some point at distance R from the source is determined by the modulus of the Umov–Pointing vector:

$$I = \frac{c}{4\pi} EH = \frac{c}{4\pi} E^2. \tag{3.8}$$

In this section, dealing with the spatial distribution of bremsstrahlung, it has been shown that at the electron velocities close to the velocity of light (β→1), the emission of the electrons propagates in the direction of movement of the electrons, i.e., along the tangent to the circular orbit of the electrons. The angular divergence of synchrotron radiation is determined by the expression:

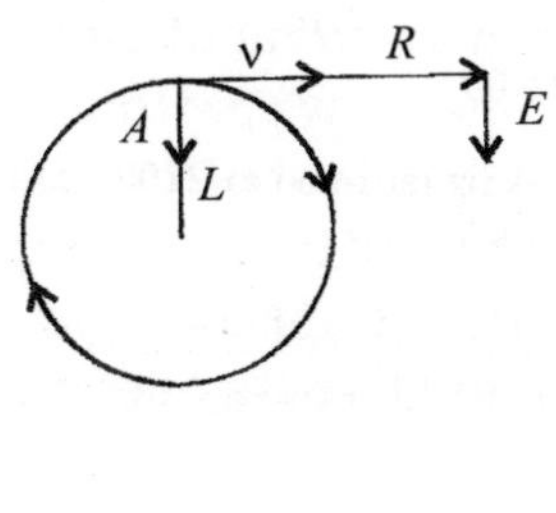

Fig 3.6. The diagram of movement of the electrons in a synchrotron and the formation of radiation at the observation point at the distance R.

$$\psi = \frac{mc^2}{\varepsilon_{el}} \ll 1,$$

where mc^2 is the rest energy of the electron, ε_{el} is the kinetic energy of the electron determined as $\frac{mc^2}{\sqrt{1-\beta^2}} - mc^2$, i.e. as the difference between the energy of the moving and resting electrons.

This dependence shows that the angular aperture of the synchrotron radiation rapidly decreases with increasing electron energy. In the x-ray energy range, this aperture is usually of the order of 0.1 mrad.

The strength of the electrical field at a distance R from the source along the tangent to the orbit (the angle φ between the vector of acceleration of the electrons and the direction of the vector $\mathbf{R}$ to the observation point is equal to $\pi/2$) is determined only by the electron energy ε_{el} and the radius L of the orbit of the electrons:

$$E \propto \frac{\varepsilon_{el}^3}{RL}. \tag{3.9}$$

In turn, the radius of the orbit L depends on electron energy ε_{el} and the strength of the magnetic field H in the synchrotron:

$$L \propto \frac{\varepsilon_{el}}{H}. \tag{3.10}$$

Substituting the values, determined by (3.9) and (3.10), into the equation (3.8) for the modulus of the Umov–Pointing vector, we obtain the following equation:

$$I \propto \frac{\varepsilon_{el}^4 H^2}{R^2}. \tag{3.11}$$

Equation (3.11) shows that the intensity of synchrotron radiation rapidly increases with increasing electron energy and the strength of the magnetic field.

Spectral distribution of synchrotron radiation in the x-ray range at the electron energy of 3.5 GeV and a current of 200 mA is shown in Fig. 3.7.

In the low-energy range of the x-ray spectrum, the intensity of radiation slowly increases in proportion to the cubic root of the quantum energy, and the short-wave range is accompanied by a more rapid exponential decrease of intensity. The maximum of the spectral distribution of synchrotron radiation corresponds to the photon energy determined by the dependence (see, for example [128])

$$E_{max} = \frac{2.218\varepsilon_{el}^3}{L},$$

where to obtain the photon energy E_{max} in keV, the energy of the electron beam ε_{el} in the synchrotron should be specified in GeV and the radius L of the orbit of the electrons in metres.

The spectral radiation in the plane of the orbit of movement of the electron beam is completely polarised. This is easily explained by the fact that the trajectories of deceleration of the electrons at the moment of emission of photons by the electrons are the same and strictly fixed.

As indicated by Fig. 3.7, synchrotron radiation is characterised by very high intensity. This high intensity of x-ray synchrotron radiation in comparison with the intensity of the x-ray tubes is explained as follows:

– as mentioned previously, only a small part of the electrons of the cathode beam decelerates with the emission of x-ray ions, whereas all the electrons, moving along the orbit, emit in the synchrotron;

– the power of the x-ray tubes is restricted by the possibility of failure of the anode under the effect of the electron beam. The current of the electron beam of the synchrotron is of the order of hundreds of milliamperes and is considerably stronger than the anode current of the x-ray tubes;

– the emission of the anode of the x-ray tube is almost completely isotropic since the synchrotron radiation is concentrated in the range of the small solid angles, so that the highest density of the flux may be produced.

The currently available synchrotrons are characterised by even higher intensity of x-radiation which exceeds the intensity of the most powerful x-ray tubes by approximately 10^{12} times. In these sources (third generation), the electrons do not move around the circular orbit of the storage ring and intersect the straight section containing a system of magnets (ondulator) which generates the oscillations of electrons with a low amplitude. The x-

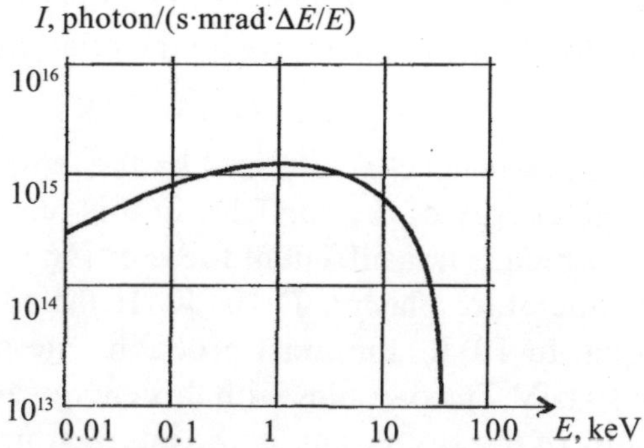

Fig. 3.7. Spectral distribution of synchrotron radiation in the x-ray range at the electron energy of 3.5 GeV and a current of 200 mA.

ray radiation, emitted during oscillations, is coherent leading to the formation of radiation whose intensity is many times higher than that observed in the movement of the electrons along the circular orbit. More detailed information on the synchrotron sources of the third generation has been published in [129].

The high intensity of synchrotron radiation and its polarisation make it possible to solve a large number of problems in practice. In particular, the excellent results obtained with its help in analytical chemistry [130, 131], in studies of the structure and properties of various substances and materials, and in the optical formation of microfocus wide-aperture x-ray beams [132], are very important.

Unfortunately, the difficulties in the construction and service of synchrotrons result in considerable restrictions of the extent of application.

3.5. X-ray radiation of high-temperature plasma

If the plasma is regarded as a statistically equilibrium system, the most probable velocity of movement of the particles is determined by the maximum of the continuous Maxwell distribution with respect to the velocities of these particles:

$$v = \sqrt{\frac{2kT}{m}},$$

where T is the temperature of matter, m is the particle mass, k is the Boltzmann constant.

The energy of the particles, corresponding to this velocity, is:

$$E = \frac{mv^2}{2} = kT.$$

If the Boltzmann constant k is defined in the units eV/K, the energy of the particles is linked with the temperature by the relationship:

$$E\,[\text{eV}] = 8.6174 \cdot 10^{-5}\, T\,[\text{K}]$$

Low-temperature plasma is characterised by the temperatures $<10^6$ K, i.e., the most probable energy of the particles in this plasma is of the order of hundreds of eV, which is not sufficient for exciting x-ray radiation.

For high-temperature plasma $T>10^6$ K. If the plasma temperature is regarded as equal to 10^8 K, the most probable energy of the particles is approximately 10 keV. The particles with this energy are capable of ionising the internal shells of the atoms, and in the interaction with the atom nuclei they generate bremsstrahlung. The main sources of x-radiation are the free electrons, formed in the high-temperature plasma. Actually, the effective ionisation of the atoms requires that the electrons have the energy which is 2–2.5 times higher than the binding energy of the electrons in the atom. The binding energy of the electrons of the K-shell of the elements with low atomic numbers represents units of keV for calcium ($Z = 20$) – 4.03 keV, for chromium ($Z = 24$) – 5.99 keV i.e., for these elements, the conditions are close to optimum. The electrons also provide a controlling contribution to the formation of bremsstrahlung of plasma because the mass of these electrons is considerably smaller than the mass of any ions (section 3.2).

The excitation of the characteristic x-irradiation by the ions is unlikely to take place because the effective ionisation of the atoms, as mentioned in section 3.2, requires the energy of the order of several MeV.

These considerations should be regarded as the lower estimate of the particle energy. Usually, the plasma is subjected to the effect of external electrical and magnetic fields, i.e., the system is not equilibrium. For the nonequilibrium system, the velocity of the particles can be determined from the Boltzmann kinetic equation. We examined the case of the presence of strong external fields found in, for example, the high-temperature plasma of laser systems working in the optical range.

The strength of the electrical field in advanced high-intensity ($I > 10^{18}$ W/cm^2) sub-nanosecond laser systems reaches values of the order of 10^8–10^{10} V/cm (see, for example [133]) which is several orders of magnitude higher than the accelerating potential in conventional radio-frequency accelerators. In such a strong field, the electrons are accelerated to relativistic velocities [134] and penetrate into the region of the cold target leaving behind the front of the thermal wave. This leads to the formation of characteric radiation and bremsstrahlung [135, 136]. The transformation coefficient of laser radiation into hard x-radiation is of the order of hundredths of a percent (according to [137]). However, taking into account the very high power of the high intensity sub-nanosecond lasers, the x-radiation during pulses is very strong. The hard bremsstrahlung at the relativistic velocities of the electrons is directed into the thickness of the target and this is in agreement with the theoretical considerations, presented in section 2.2.

The spectral composition of the x-radiation of the high-temperature plasma in radio-frequency accelerators (Tokamak and other similar sources) is characterised by several intermediate frequencies between the extreme cases of plasma radiation, examined here. The experimental data confirm that in these cases bremsstrahlung and characteristic K-radiation of certain elements (for example, CaK- and FeK-lines) are generated.

The thermal process is one of the processes forming the x-ray radiation of the Sun and other stars.

3.6. X-ray lasers

Generation of laser radiation requires the inverse state of matter in which the number of the exciting particles is greater than the number of the non-excited ones. The specific power, required for maintaining the inverse state, is determined by the dependence (see [138]):

$$W \sim \nu^4 = \text{cont} \cdot \lambda^{-4},$$

i.e., the mean power is inversely proportional to the fourth power of the wavelength of the generated laser radiation. If the specific power of the order of 10^2–10^4 W/cm^3 is sufficient for the visible range ($\lambda \approx 500$ nm), then the required specific power for the x-ray range ($\lambda \approx 0.5$ nm) is 10^{14}–10^{16} W/cm^3.

This power is at present found in nuclear explosions or in the focus of a powerful laser source, working in the pulsed regime.

The literature (for example, [139]) contains reports on the successful realisation of x-ray lasers with pumping by nuclear explosion.

At the focus of a laser with a specific power of the order of 10^{17}–10^{18} W/cm^3 the matter is completely ionised and converted to the plasma state in which the resultant ions are almost completely free from the electron shell. Losing part of the kinetic energy in the cooling process, the electrons are trapped by the ions and plasma recombination takes place. The probability of capture of the electrons by higher energy levels is higher than that by the lower levels and this results in the formation of the inverse state and enables the medium to intensify radiation with the frequency corresponding to the difference of the energies of the internal and external shells of the atom. This process of formation of the inverse state is referred to as recombination.

The first successful laboratory experiments with the construction of an x-ray laser in the plasma of multicharged ions were carried out in 1984 [140–142].

Figure 3.8 shows the possible mechanism of heating the plasma at the focus of an optical laser, published in [138]. Laser radiation 1 is focused by the cylindrical lens 2 on the target 3 in the form of a longitudinal spot whose length D and width h are close to $h \approx 10\lambda$. Assuming that the depth of penetration of radiation into the plasma is 10λ, the wavelength of laser radiation $\lambda = 1$ µm and $D \approx 1$ cm, the volume of the plasma is of the order of 10^{-6} cm^3. At the energy of the laser radiation pulse of 1 J and the pulse time 10^{-9} s, we obtain the power of the order of 10^9 W and the specific power of the order of 10^{15} W/cm^3.

This value is in agreement with the previously mentioned specific power, required for maintaining the inverse state in the x-ray range. However, in

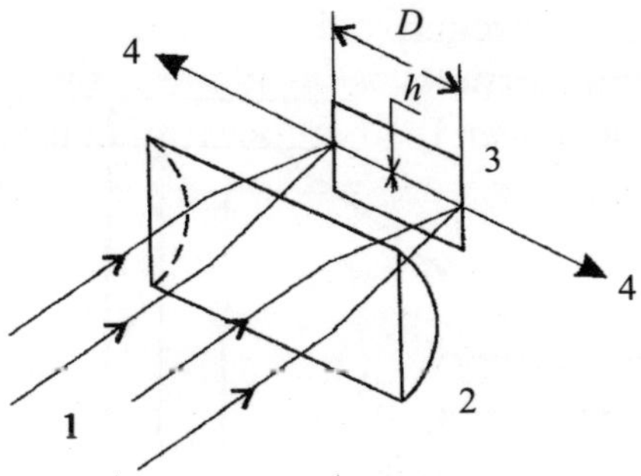

Fig. 3.8. Laser heating of plasma. 1) laser radiation; 2) cylindrical lens; 3) metallic target; 4) x-ray radiation. *D* and *h* are the longitudinal and transverse dimensions of the plasma.

the conditions of actual experiments, it is necessary to have a large power reserve.

In experiments carried out and the Lawrence National Laboratory of the USA in the equipment whose principal diagram is similar to that shown in Fig. 3.8, the radiation in the range of the wavelength of the order of 4–7 nm in laser heating of targets of different metals was greatly intensified [143].

In the x-ray lasers with plasma heating with radiation in the range of tens of nanometres, there are transitions between the discrete levels of the excited states with the filled internal shells of the plasma ions. As an example, we examine the scheme of the $3p$–$3s$ transitions in neon-like ions $1s^22s^22p^53p$ (J = 2)$\to 1s^22s^22p^53s$ (J = 1), published in [144].

The inverse population for $3p \to 3s$ transitions using the quasi-stationary plasma of nanosecond equipment is generated on the three-level pumping scheme (Fig. 3.9). The level J = 2 is pumped as a result of the excitation of the ground state of Ne-like ions (configuration $1s^22s^22p^6$).

The transition of the electrons from these levels to the ground state, where J = 0, is not permitted by the rules of selection and is strongly suppressed. However, the level J = 1 is 'cleaned' as a result of transitions to the ground state at a considerably higher rate than the rate of filling from this state under the effect of the free plasma electrons. Consequently, we obtain the inverse population of the investigated levels. The wavelength λ of laser radiation is determined by the difference of the energies of $3p$ and $3s$ levels. For neon-like selenium ions it appears that $\lambda = 20.6$ nm and $\lambda = 20.9$ nm.

A number of other physical mechanisms of pumping of x-ray lasers have been examined in, for example, [145].

The application of picosecond and femtosecond pulses has greatly reduced the pumping energy of the x-ray lasers and helped to construct compact laser systems.

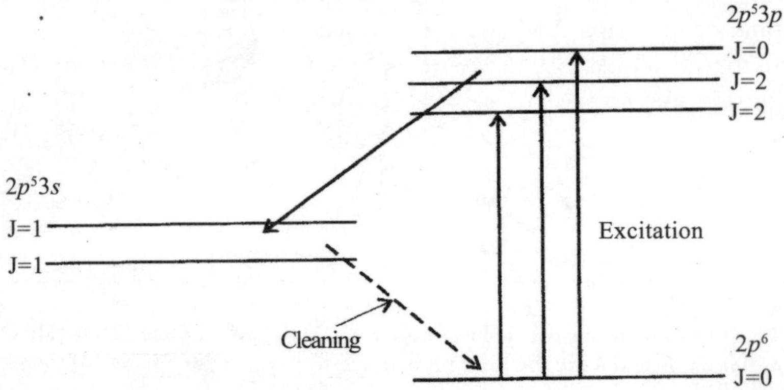

Fig. 3.9. Energy levels of the x-ray laser, working on the $3p \rightarrow 3s$ transitions of a neon-like ion.

Another principal possibility of the construction of x-ray lasers is associated with the application of non-linear optical effects [146]. In optical quantum generators, the effect of a strong electromagnetic field with a frequency ω_0 on the medium causes that the medium emits waves with the frequencies divisible by ω_0. In this case, it is not essential to generate the inverse state in the internal shells of the atom.

An essential condition for the pumping of energy from the initial frequency to the short-wave harmonics is the transparency of the medium. No solid-state substances are transparent in the short-wave range. Therefore, gas media have been used as non-linear media for these applications.

The laser systems with picosecond pulses (10^{-12} s) and the power of up to 10^{12} W produce radiation at a frequency of higher harmonics up to the 141st which corresponds to the wavelength λ of the order 7.5 nm [138].

Finally, free electron lasers working in the x-ray region of the spectrum were constructed in the 90s of the previous century [147, 148]. Clusters of electrons are accelerated in a linear accelerator to relativistic velocities and incident on the transverse periodic field of the magnetic ondulator. The density of the cluster in the linear accelerator is many times higher than the density of the blob of the electrons obtained in the synchrotron. Without this increase of the density it is not possible to induce the process of self-amplifying stimulated emission (SASE) which is the basis of the formation of laser radiation. The SASE increases the produced power by several orders of magnitude. The zigzag acceleration of the electrons in the field of the ondulator results in the formation of magnetic bremsstrahlung propagating in the direction of movement of the electrons. For small wave effects of the ondulator, radiation is quasi-monochromatic. In each cycle of the ondulator, the radiation phase outstrips the movement of the electron by the resonance wavelength. The mechanism of self-amplified stimulated emission makes it possible to increase the emitted power by several orders of magnitude. At present, the wavelengths of the order of hundreds of nanometres have been achieved for the laser radiation of free electrons with prospects of moving into the region of shorter wavelengths

Chapter 4

Absorption of x-ray radiation

Absorption is one of the most important processes leading to attenuation of the x-ray radiation flux during its interaction with matter. In addition to absorption, the weakening of x-ray beams is determined by the scattering of the beam, and for the shortwave photon radiation (with the quantum energy higher than 1 MeV) by the formation of the electron–positron pair in the field of the atomic nucleus and of a triplet during interaction with the atomic electrons. The scattering of x-ray radiation will be investigated in Chapter 5. The last two processes do not relate to the physics of x-ray radiation because the range of application of the radiation is restricted to energies lower than 150 keV.

4.1. Electronic, partial and atomic coefficients of absorption

The probability of absorption of x-ray radiation by a single electron of the atom shell will be determined. This probability is characterised by the electronic absorption coefficient τ_e and is determined as the ratio of the energy W, adsorbed by the electron per unit time t, to the intensity I of x-ray radiation (the energy incident on the unit area in unit time):

$$\tau_e = \frac{W/t}{I}. \quad (4.1)$$

Since the intensity I is the energy incident in unit time on the unit area, the dimension of the electronic coefficient of absorption in equation (4.1) is $[\tau_e] = cm^2$. This means that the electronic coefficient of absorption is the probability of absorption of the beam of x-ray radiation with the area of 1 cm^2 by a single electron of the atom, situated within this cross-section. In other words, τ_e is the cross-section of interaction of x-ray radiation with the electron of the atom.

The equation for the electronic coefficient of adsorption will be derived on the basis of the classic theory which treats the electron of the atom as some oscillator [7].

The intensity I in equation (4.1) is determined by the modulus of the Umov–Pointing vector:

$$I = \frac{c}{4\pi}\overline{E}^2, \tag{4.2}$$

where $\overline{E}$ is the average strength of the electrical field in the region of the oscillating electron.

Quantity W/t for equation (4.1) is determined by examining the classic electronic oscillator in which the atomic electron is maintained in the equilibrium position by the force:

$$F(x) = -4\pi^2 m\nu_q^2 x, \tag{4.3}$$

where m is the electron mass, ν_q is the intrinsic frequency of the oscillations of the electron.

If under the effect of x-ray radiation the electron is deflected to the distance dx from the equilibrium position, the work of the decelerating force $F(x)$ in extinction of the forced oscillations of the electron, determined by the effect of the external magnetic field, is determined by the following equation:

$$W = \int_x F(x)dx. \tag{4.4}$$

This work is equal to the required energy absorbed by the electron.

We determine the value of W for the single period T of forced oscillations:

$$W_T = \int_0^{x_T} F(x)dx = 2\pi^2 m\nu_q^2 x_T^2. \tag{4.5}$$

The path x_T, travelled by the electron in half the period of forced oscillations and of the effect of some mean strength $\overline{E}$ of the electric field, is determined from a simple kinetic relationship:

$$x_{T/2} = \frac{\overline{a}}{2}\left(\frac{T}{2}\right)^2, \tag{4.6}$$

where the average value of acceleration $\overline{a}$ is determined by the equation:

$$\overline{a} = \frac{e\overline{E}}{m}.$$

Taking this equation into account and accepting that $T=\lambda/c$, we obtain:

$$x_{T/2} = \frac{e\bar{E}}{8m}\frac{\lambda^2}{c^2}.$$

In the second half of the period under the effect of the electrical field with the opposite sign of the strength, the decelerating electron may travel exactly the same distance to complete rest at the end of the period. Therefore, the path travelled by the electron during the entire period is determined as follows:

$$x_T = \frac{e\bar{E}}{4m}\frac{\lambda^2}{c^2}. \tag{4.7}$$

Substituting the resultant value into the expression (4.5) for the energy absorbed by the electron, we obtain:

$$W_T = \frac{\pi^2}{8}\nu_q^2\frac{e^2(\bar{E})^2}{m}\frac{\lambda^4}{c^4}.$$

In order to determine the energy W_T, adsorbed in unit time, the resultant expression is divided by the period $T=\lambda/c$. We obtain:

$$\frac{W_T}{T} = \frac{\pi^2 e^2}{8mc^3}(\bar{E})^2\lambda^3\nu_q^2. \tag{4.8}$$

In equation (4.8) we determine the frequency ν_q of intrinsic oscillations of the electron, using the approximate value of the main term for the energy of the q-level of the atom in the equation (2.1):

$$\varepsilon_q = h\nu_q = Rhc\frac{(Z-\sigma_1)^2}{n^2},$$

where n is the main quantum number for the q-level.

For the K-shell, the complete screening constant σ_1 is equal to 0.3 and, therefore, in the case of relatively high values of Z the constant may be ignored. For other shells, in which the value of σ_1 is quite high, neglecting the constant may result in large errors.

Nevertheless, we may write $\nu_q=RcZ^2/n^2$. Substituting the resultant expression into equation (4.8) gives:

$$\frac{W_T}{T} = \frac{\pi^2 e^2}{8mc}\frac{R^2}{n^4}(\bar{E})^2\lambda^3 Z^4. \tag{4.9}$$

Knowing the value of the energy adsorbed by the electron in unit time and the intensity of the incident x-radiation, determined by the expression (4.2), on the basis of the equation (4.1) for the electronic coefficient of true absorption we obtain:

$$\tau_e = \frac{W_T/T}{I} = \frac{\pi^2 e^2}{2mc^2}\frac{R^2}{n^4}\frac{(\bar{E})^2}{\bar{E}^2}Z^4\lambda^3. \tag{4.10}$$

The value $(\overline{E^2})/\overline{E}^2 = 8/\pi^2$ is its constant value. Therefore, equation (4.10) is presented in the new form:

$$\tau_e = C\lambda^3 Z^4, \tag{4.11}$$

where $C = \frac{4\pi e^2}{mc^2}\frac{R^2}{n^4}$.

Equation (4.11) shows that the electronic coefficient of absorption is proportional to the cube of the adsorbed wavelength and the fourth power of the atomic number of the absorbing element.

The partial coefficient of true absorption τ_q for some shell q is determined by the same equation (4.11) multiplied by the number of electrons n in this shell, i.e., the constant has the value nC.

For the atomic coefficient of true absorption τ_a at $\lambda<\lambda_K$ we obtain:

$$\tau_a = \tau_K + \tau_{L1} + \tau_{L2} + ... = (C_K + C_{L1} + C_{L2} + ...)\lambda^3 Z^4,$$

i.e.

$$\tau_a = C_a \lambda^3 Z^4. \tag{4.12}$$

Attention should be given to the fact that the atomic coefficient of absorption, like the electronic and partial coefficients, has the dimension $[\tau_a] = cm^2$. In other words, the atomic coefficient of absorption is the cross-section in which the x-ray radiation is unavoidably absorbed by the atom.

The investigated kinetic theory of absorption of x-ray radiation is approximate. It gives mainly the accurate dependence of the atomic coefficient of absorption on the wavelength and the atomic number of the absorbing element. However, the value of the constant C_a is too high in this case.

This too high value is caused by the re-radiation by the oscillator of the energy adsorbed by it. This is not taken into account in the investigated approximation. Actually, transferring to the state of forced oscillations, the electronic oscillator starts its self to emit, returning part of the absorbed energy to the electromagnetic field.

One of the first variants of the quantum theory of absorption of x-irradiation was proposed by Stobbe [149] who used the Einstein equations [150] linking the probabilities of absorption and emission of radiation during transition between the energy levels of the atom. In the process of conversion of these equations, Stobbe introduced a number of significant simplifications. The results obtained by Stobbe indicated that the power dependence of the partial coefficient of absorption on the wavelength λ and the atomic number Z is more complicated in comparison with the dependence predicted by the classic theory. The powers at λ and Z for the partial coefficient of absorption are not integers. In addition to this, it has been shown that the wavelength of absorbing radiation has a significant effect on the relationship between the partial absorption coefficients. For example, for the L-shell it was obtained that:

$$\frac{\tau_{L1}}{\tau_{L2} + \tau_{L3}} = \frac{1}{6}\frac{\lambda_L}{\lambda}, \qquad (4.13)$$

which means that at the wavelengths λ close to the wavelength λ_L, corresponding to the binding energy of the electrons in the L-shell, the investigated ratio of the partial absorption coefficients is close to 1:5.5. If $\lambda = \lambda_L/6$, this ratio becomes 1:1. Thus, the short-wave radiation is efficiently absorbed by the L_1-sublevel, and the long-wave radiation is absorbed mainly by the L_2- and L_3-sublevels of the atom. This conclusion is in agreement with the results of investigations in [151] of the magnetic spectra of the photoelectrons which show the increase of the ratio of the number of the L_3-photoelectrons to the number of L_1-photoelectrons on approach of the wavelength of the absorbing radiation to the L-edge of absorption.

Calculations carried out on the basis of the Stobbe theory provide very approximate values of the absorption characteristics. Therefore, semiempirical variants of the determination of the atomic cross-sections of photoelectric absorption were proposed. They use the detailed theory of determination of the scattering cross-section, confirmed by the experimental data. Photoelectric absorption was determined as the difference between the experimentally determined attenuation of x-ray radiation and the calculated values of coherent and non-coherent scattering of radiation (see, for example [152, 153]).

A more developed non-relativistic quantum theory of absorption was proposed by Scofield [154] who examined the interaction of the photon with the orbital electron, moving in the field of the Hartree–Slater static central potential. Slightly later, the ionisation cross-section was calculated [155], using the Dirac–Hartree–Fock relativistic theory of the self-consistent field. The accuracy of these theoretical calculations of the photoionisation cross-section is estimated [156] by the value of the order of 2–5% for the photon energy from 1 keV to 10 MeV. At lower quantum energies, the accuracy of the calculations rapidly decreases. The accuracy also decreases for the photon energy close to the energy of the edges of absorption of the elements. This is caused by the large variation of the probability of absorption of photons at the energies close to the binding energy of the electrons in the ionised shell.

The tabulated data available in [155–157] on the atomic and partial cross sections of photoionisation, calculated using the theory of the self-consistent field, are used widely in computational modelling of the absorption process in many practical applications.

According to the current views [129], the process of absorption is regarded as an interaction between the incident photon and the atomic electron, using the interaction Hamiltonian H_i which characterises the transition of the system from the initial state i to the final state f. The initial state $|i\rangle$ is determined by the state of the incident x-ray photon and the atomic electron with which the proton interacts $|i\rangle=|1\rangle_p|0\rangle_e$. The final state $|f\rangle$ is given as $|f\rangle=|0\rangle_p|1\rangle_e$, where the photon was annihilated and the photoelectron was removed from the atom. The interaction Hamiltonian for the absorption process in the simplest case of the non-bonded final state of the photoelectron is determined by the expression:

$$H_i = \frac{eAp}{m},$$

where p is the moment of the electron, A is the vector potential.

In a general case, we obtain a complicated equation for the interaction process which determines the number of cases dW/dt of absorption of the x-ray photons by some shell of the atom in unit time. After several simplifications we obtain a simple dependence for the partial coefficient of adsorption of the x-ray photons by the K-shell:

$$\tau_K = 32\lambda r_0 \left(\frac{4}{3}\right)\left(\frac{\lambda}{\lambda_L}\right)^{5/2} f(\xi),$$

where r_0 is the Thomson scattering length (see chapter 5); $\xi \approx \sqrt{\frac{\lambda}{\lambda_K - \lambda}}$ is a variable introduced by Stobbe [149].

The quantity $f(\xi)\to 1$ at $\lambda \ll \lambda_K$ and $f(\xi)\to 2\pi/e^4$ at $\lambda\to\lambda_K$.

4.2. Absorption jumps

The absorption jumps are determined by the process of photoelectric interaction of the photons with the atoms of matter.

If the energy of the excess photon is not sufficient for the ionisation of the K-shell of the atom ($\lambda_K<\lambda<\lambda_{L1}$), then this shell does not take part in the investigated process and the following equation should be accepted for the atomic coefficient of absorption:

$$\tau'_a = \tau_{L1} + \tau_{L2} + \ldots \tag{4.14}$$

Thus, at $\lambda>\lambda_K$ the absorption coefficient decreases by the value τ_K. According to the equations (4.12) and (4.14), the resultant absorption jump S_K is:

$$S_K = \frac{\tau_a}{\tau'_a} = \frac{\tau_K + \tau_{L1} + \tau_{L2} + \ldots}{\tau_{L1} + \tau_{L2} + \ldots} = \frac{\tau_K}{\tau'_a} + 1,$$

and consequently

$$\tau_K = (S_K - 1)\tau'_a,$$

or

$$\tau_K = \frac{(S_K - 1)}{S_K}\tau_a. \tag{4.15}$$

The equation (4.15) links the partial coefficient τ_K of absorption of radiation by the K-shell with the atomic coefficient of absorption for the wavelength which is shorter than the wavelength λ_K of the investigated absorption edge. It is assumed that τ_K and τ_a are proportional to $\lambda^3 Z^4$.

Figure 4.1 shows the dependence of the atomic coefficient of absorption of platinum on the wavelength of absorbing radiation. The absorption coefficient rapidly increases with the increase of the wavelength of the absorbing radiation (see equation (4.12)) and the coefficient rapidly decreases at the wavelength corresponding to the K-edge of absorption (for platinum $\lambda_K = 0.158$ Å) and then at the wavelength of L_1-, L_2- and L_3-edges of absorption.

The dependence of the atomic coefficient of absorption on the atomic number of the absorbing element is also characterised by the presence of absorption jumps, as the dependence of τ_a on the wavelength, shown in Fig. 4.1. In fact, at constant wavelength of adsorbed radiation, this coefficient increases in proportion to the fourth power of the atomic number. However, at some value of Z, the energy of the quanta of x-ray radiation becomes insufficient for the ionisation of the K-shell of the atom. This shell is excluded from the absorption process and the absorption coefficient rapidly decreases. A K-jump is found. With a further increase of Z, L_1- , L_2- and L_3-absorption jumps are detected.

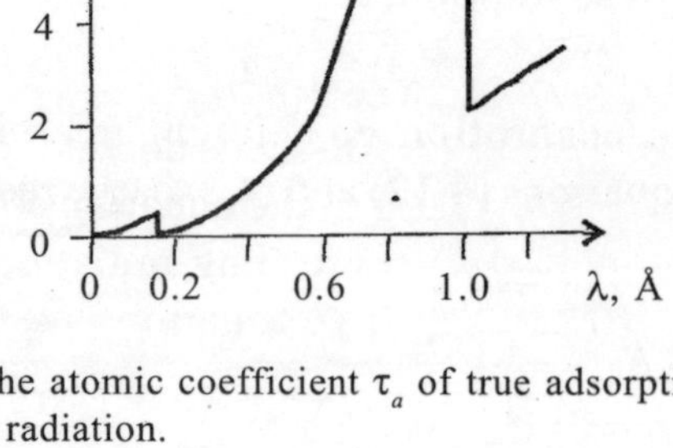

Fig. 4.1. Dependence of the atomic coefficient τ_a of true adsorption of platinum on the wavelength λ of absorbing radiation.

For the partial absorption coefficients of L_1-, L_2-, L_3-, etc. shells, the equations are identical to the equation (4.15), if there are no other edges of absorption of the given element between the examined wavelength λ and the wavelength of the appropriate absorption edge, i.e., if the following conditions are fulfilled:

$$\tau_{L1} = \frac{S_{L1}-1}{S_{L1}}(\tau_a)_\lambda \quad \text{at} \quad \lambda_K < \lambda < \lambda_{L1},$$
$$\tau_{L2} = \frac{S_{L2}-1}{S_{L2}}(\tau_a)_\lambda \quad \text{at} \quad \lambda_{L1} < \lambda < \lambda_{L2}, \qquad (4.16)$$

and so on.

If some other absorption edge of this element is found between the wavelength λ and the examined absorption edge, it is necessary to take into account the additional jump at the wavelength of this absorption edge. For example, if $\lambda<\lambda_K$, we obtain the following equation for the partial coefficient of absorption of radiation by the L_1-sublevel:

$$\tau_{L1} = \frac{S_{L1}-1}{S_K S_{L1}}(\tau_a)_{\lambda<\lambda_K}. \qquad (4.17)$$

Similarly, for τ_{L2} at $\lambda<\lambda_K$ we obtain:

$$\tau_{L2} = \frac{S_{L2}-1}{S_K S_{L1} S_{L2}}(\tau_a)_{\lambda<\lambda_K} \qquad (4.18)$$

and so on.

It should be mentioned that these equations, like the dependences (4.15) and (4.16), are approximate. They are based on classic considerations regarding the absorption of x-ray radiation which does not take into account the complicated dependence of the partial coefficient of absorption on the atomic number Z and on the wavelength λ of absorbing radiation, established at a later date. The results of conversion of the total absorption coefficients to partial coefficients using the equations (4.15)–(4.18), are compared in [32] with the calculated data obtained by Schofield [154], using the Hartree–Slater theory. It was found that the error of application of equation (4.15) for the K-level is relatively small and amounts to 1–3%. However, the ratio of the total and partial absorption coefficients for the individual sublevels of the L- and M-shells does not correspond to the classic considerations, forming the base of the equations (4.16)–(4.18). This ratio depends on the wavelength of absorbing radiation and this confirms the conclusions of Stobbe' s theory (see equation (4.13)) and may change by a factor of 2–4 with the variation of the wavelength. Regardless of this, the simple form of these equations is the reason why they are used widely, for example, in the calculation of the intensity of x-ray fluorescence.

To determine the common absorption jump of the entire L-series, detected in transition from the wavelength range $\lambda_K<\lambda<\lambda_{L1}$ to the range $\lambda_{L3}>\lambda>\lambda_{M1}$, it is necessary to use the partial absorption coefficient τ_L which is determined as the sum of τ_{L1}, τ_{L2} and τ_{L3} partial absorption coefficients:

$$\tau_L = \tau_{L1} + \tau_{L2} + \tau_{L3} = \left(\frac{S_{L1}-1}{S_{L1}} + \frac{S_{L2}-1}{S_{L1}S_{L2}} + \frac{S_{L3}-1}{S_{L1}S_{L2}S_{L3}}\right)(\tau_a)_{\lambda_K<\lambda<\lambda_{L1}}.$$

Consequently

$$\tau_L = \frac{S_{L1}S_{L2}S_{L3}-1}{S_{L1}S_{L2}S_{L3}}(\tau_a)_{\lambda_K<\lambda<\lambda_{L1}}. \qquad (4.19)$$

Equation (4.19) shows that the absorption jump S_L for the entire L-series is determined as the product of the partial absorption jumps of the individual sublevels:

$$S_L = S_{L1}S_{L2}S_{L3}.$$

The values of the K-, L_3- and M_5-absorption jumps for several elements are presented in Table 4.1. The absorption jumps for other sublevels of the L- and M-series are usually assumed to be constant [33]:

$$S_{L1}=1.2,\ \ S_{L2}=1.4,\ \ S_{M1}=1.07,\ \ S_{M2}=1.11,$$
$$S_{M3}=1.32,\ \text{and}\ S_{M4}=1.32.$$

Table 4.1 shows that the K-absorption jump is more significant. The magnitude of the absorption jumps decreases with the increase of the atomic number Z of the absorbing element.

4.3. Structure of absorption edges

The electrons, injected during absorption from some inner atomic level, may transfer to some other outer, empty level of this atom or to some other unfilled level common for the molecules, and also in the continuous range of the possible energy values. If the photon energy does not correspond to the energy of the unfilled level or the absorption transition is not permitted by the dipole selection rules, this transition is not possible.

The transition of the electrons to the continuous energy range was investigated in the monographs [7] where this range is regarded as the set of some resonances levels f close to each other. The inner level of the atom has some natural width. Therefore, the transition of the electron from this level at its energy E_i to one of the resonance levels with the energy E_f takes place in the manner shown in Fig. 4.2, where $p(E)$ is the probability of disposition of the i-electron energy in the limits from E_i to E_i+dE_i.

Table 4.1. Values of the absorption jumps for K-, L_3- and M_5-levels of atoms of several elements

Element (atomic number)	S_K	S_{L3}	S_{M5}
F (9)	17.4	–	–
Al (13)	13.1	7.66	–
Ca (20)	9.75	5.50	–
Cu (29)	7.82	4.26	–
Mo (42)	6.48	3.41	–
W (74)	5.19	2.58	2.69
U (92)	4.86	2.37	2.10

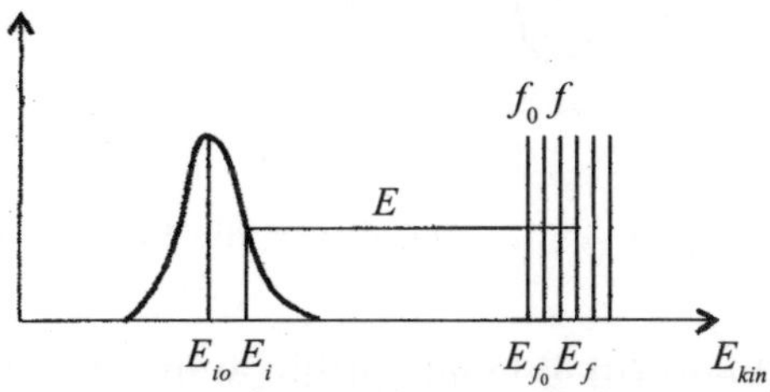

Fig. 4.2. Adsorption of a photon with energy E by resonance level f. E_i is the binding energy on the level i taking into account its natural width, E_f is the energy of the resonance level ι.

The energy E of the absorbed photon is determined from the equation:

$$E = E_i + E_f.$$

The values of the coefficients of absorption τ_f by the resonance level f for the photons, knocking out the electrons from the level i, are distributed on the energy scale in accordance with the probabilities of distribution of the energie E_i of the electrons of the i-level within the limits of the natural width of the level.

If it is assumed that the transitions to all resonance levels from the level i are equally probable, we obtain a set of identical distributions. The set starts with some boundary value corresponding to the minimum energy E_{f0} of the resonance levels (Fig. 4.3). Assuming the dispersion nature of the distribution for the individual resonance levels, after integration of these levels with respect to the energy we obtain the arctangent shape of the absorption jump shown in the figure which in the previous examination was regarded as the final jump-like break of some continuous function.

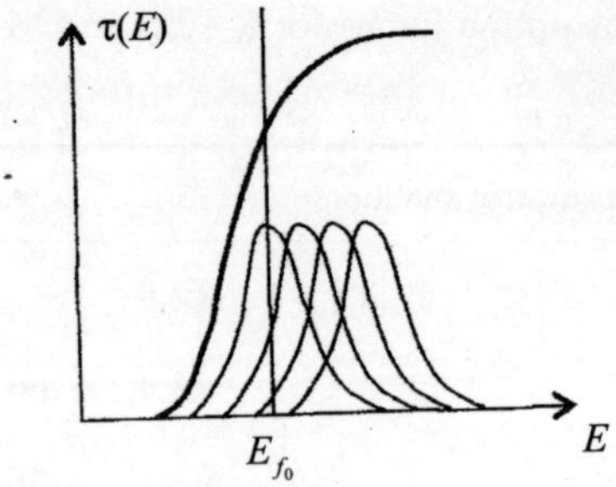

Fig. 4.3. Total absorption coefficient in transition of the electrons from the internal *i*-level to the continuous energy range.

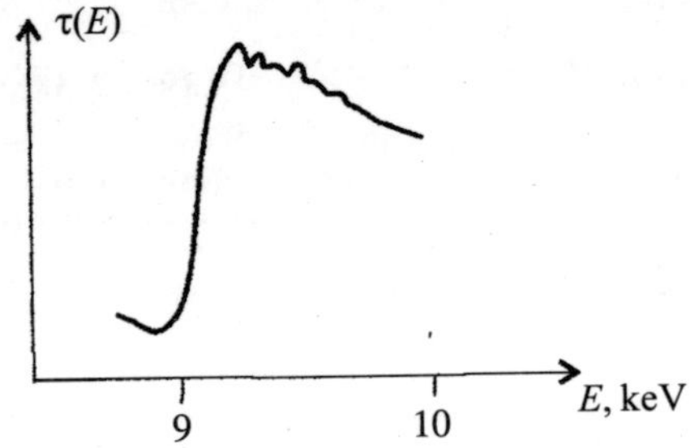

Fig. 4.4. The structure of the absorption edge of copper.

For the metals, the unfilled levels of the atoms appear to be connected with the conduction band. During absorption, the electron is injected behind the Fermi boundary. The transitions of the electrons to the resonance levels in the continuous conductivity band are not equally probable. Therefore, after the arctangent jump, the absorption coefficient fluctuates. These fluctuations take place in the range 400–500 eV in the region of higher energies in relation to the absorption edge. As an example, Fig. 4.4 shows the experimental spectrum of absorption of copper, published in [158].

The density of the resonance levels in the investigated case is determined by the nearest environment of the absorbing atom. Therefore, the variations of the absorption coefficient, shown in Fig. 4.4, may be interpreted as fluctuations, determined by the interference of the spherical wave function of the photoelectron with its part, inversely scattered by the atoms of this environment [129]. This process may be described using the interaction Hamiltonian, investigated in the previous section. The initial state of the system (photon + atomic electron) remains unchanged, and the final state of the system (photoelectron) differs by the fact that the wave function f_0 of the free photoelectron is corrected by some value Δf, which takes into account the inverse scattering of the spherical wave of the photoelectron and its interference with the emitted wave:

$$f \rightarrow | f_0 + \Delta f \rangle.$$

The interaction Hamiltonian should take into account not only the spherical divergences of the wave of the photoelectron and the inversely scattered wave, but also the phase shift, caused by the electrostatic potential between the negatively charged electron and the ions of the lattice. In addition, it should take into account the oscillation of the adjacent atoms around their equilibrium position in the lattice, and also the fact that the adjacent atoms should be of different type and their distance from the absorbing atom should differ. Consequently, we obtain a complicated equation, which may be used to calculate the value of the fluctuation of the absorption coefficient on the condition that the structure of the lattice is given, and also calculate the distances, number and scattering capacity of the adjacent atoms.

Of considerable interest is the solution of the inverse problem of determination of the nearest environment of the absorbing atom on the basis of the value of the previously mentioned fluctuations of the absorption coefficient. This problem is solved using the EXAFS method (extended x-ray absorption fine structure). A priori information on the possible nearest environment of the absorbing atom, and also the selection of the number, type of adjacent atoms and their position in the lattice make it possible, by successive approximations, to reach agreement between the calculated and measured fluctuations of the absorption coefficient. The final parameters of the nearest environment, used in the calculations, are regarded as true values.

We now examined the case in which the unfilled valency atom shells do not merge with the region of continuous energies. In this case, the absorption cross-sections whose absorption coefficient is greater than the jump, shown in Fig. 4.4, appear closely to the investigated absorption edge. This jump is known as the 'white line' which appeared in the recording of the spectrum on photographic film in early x-ray experiments. The simplest example of the formation of these lines is the case of monoatomic gases. One of the first attempts to explain the structure of the absorption edge by the monatomic gases was made in [159] where rules were proposed for the selection of intra-atomic transitions, similar to those existing for the emission spectrum. The main rule is the condition of the variation of the azimuthal quantum number l by unity ($\Delta l = 1$). As an example, we examine the K-spectrum of absorption of the monatomic gas, i.e., argon. In this case, the electron is removed from the $1s$-shell during absorption. According to the selection rules, $\Delta l = 1$. Consequently, transitions are possible only on the p-levels of the atom. The first unfilled p-level in argon is the $4p$-level. This means that transitions are possible on $4p$-, $5p$-, $6p$-, etc levels, and the absorption spectrum is determined by the superposition of the selective lines of adsorption 1*s*→4*p*, 1*s*→5*p*, 1*s*→6*p*, and also by the previously investigated continuous absorption band.

The intensities of the selective absorption lines of the monoatomic atoms are determined by the equation:

$$\tau_n = \frac{\eta^5 \nu}{Z^7 (n^*)^2}, \qquad (4.20)$$

where n^* is the effective main quantum number of the external *np*-level; η is the effective charge of the atomic residue for this external level after ionisation of the internal K-level; ν is the frequency of the absorption lines.

The effective main quantum number n^* of the external levels is determined by the difference $n^* = n - \delta$, where n is the main quantum number of these levels, and δ is the Rydberg correction (the quantum effect).

Assuming that for all selective absorption lines $\nu = \text{const}$, $\delta = \text{const}$ and $\eta = 1$, according to equation (4.20) we obtain the following equation for the ratio of the intensities of the initial lines:

$$\tau_4 : \tau_5 : \tau_6 : \tau_7 = \frac{1}{(4-\delta)^2} : \frac{1}{(5-\delta)^2} : \frac{1}{(6-\delta)^2} : \frac{1}{(7-\delta)^2}. \qquad (4.21)$$

The value of the quantum defect for argon is known: $\delta = 1.77$. Consequently, in accordance with (4.26):

$$\tau_4 : \tau_5 : \tau_6 : \tau_7 = 1 : 0.33 : 0.15 : 0.078.$$

The resultant theoretical relationship is in good agreement with the following, experimentally determined ratio:

$$\tau_4 : \tau_5 : \tau_6 : \tau_7 = 1 : 0.31 : 0.18 : 0.085.$$

The structure of the K-edge of absorption of argon, measured on a two-crystal spectrometer [160], is shown in Fig. 4.5.

The investigated approach represents the first approximation in the

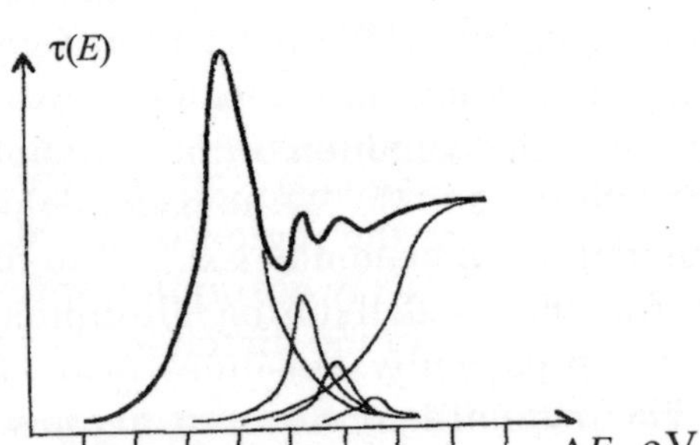

Fig. 4.5. The structure of the absorption edge of argon ($Z = 18$).

description of the fine structure of the x-ray absorption spectra.

The absolute values of the absorption coefficient $\tau(E)$ in the fine structure of the investigated K-absorption edge of argon were calculated using (see [161, 162]) Hartree–Fock many-electron wave functions. This resulted in a satisfactory agreement with the absolute experimental values of its coefficient which supplemented data on the relative intensities of the absorption peaks, given by the relationship (4.21).

For the diatomic gases, the molecular bonds distort the energy of the outer levels. In addition, transitions of the electrons to molecular orbits, which unite the external orbitals of the atoms, included in the composition of the molecule, become possible. The structure of the K- and L-edges of absorption of relatively simple molecules of HCl and H_2S, and also their bond with the structure of these molecules, were investigated in, for example, [163, 164].

In quantum mechanics (see, for example, [165]) the energy state of the unfilled external shells, which depends greatly on the coordinates of all electrons and nuclei forming the system (atom, molecule, crystal), was determined.

For the atoms and ions, the single-electron atomic orbitals may be determined by the solution of the single-electron Schrödinger equations by the Hartree or Hartree–Fock method, in which the total effective potential has spherical symmetry. For molecules searching for molecular orbitals, we examine the linear combination of the atomic orbitals, centred on the individual atoms, forming the molecule. The molecular orbital φ is expressed in the form of the sum of the individual atomic valency orbitals ψ_i with some coefficients k_i, which determines the contribution of each orbital:

$$\varphi = \sum_{i=1}^{n} k_i \psi_i, \tag{4.22}$$

where n is the number of atoms in the system.

The representation of the wave function of the individual electrons in the form of such a sum indicate that the electron with the probability given by the constant $|k_i|^2$ may be situated around every atom where it moves as a conventional valency electron.

Complex compounds have their characteristic spectral distributions of the adsorption coefficient τ_q in the region of the absorption jump. The examination of the structure of x-ray absorption spectra in the vicinity of the absorption edge is grouped in the direction known as XANES (x-ray absorption near-edge structure).

The structure of the absorption edges is an important source of information on the nature of interatomic and intermolecular bonds, the atomic spacing, the state of the crystal lattice, and on the distribution of electronic densities and other characteristics which determine the properties of matter and materials.

The interest in the spectral distribution of the absorption coefficient in the region of the edges continuously increases. The previously examined directions EXAFS and XANES are improved and the results are used on an increasing scale in advanced materials science.

4.4. Linear and mass coefficient of attenuation of x-ray radiation

In addition to the previously examined true (photoelectric) absorption, the atoms of matter scatter x-ray radiation. The overall effect of these processes determines the attenuation of the x-ray beam in materials. Correspondingly, the atomic coefficient of weakening of x-radiation, μ_a, may be described by the sum:

$$\mu_a = \tau_a + \sigma_a.$$

where σ_a is the atomic coefficient of scattering of x-radiation.

In the previous section it was shown that the coefficient of true absorption, τ_a, rapidly increases with increasing wavelength of the absorbing radiation and the atomic number of the absorbing element. The scattering coefficient does not change greatly. Therefore, in the case of the wavelengths greater than 1 Å and the atomic numbers $Z > 20$, the scattering of x-radiation may be ignored and it may be assumed that $\mu_a \approx \tau_a$. In other cases, when examining the attenuation of the x-ray beam in matter, the scattering process should be taken into account.

The variation of the intensity of the beam of x-adiation in the specimen along the path dx is determined by the intensity of radiation I_x in the depth x, by the number of atoms n in the unit volume of matter and by the probability of interaction of this radiation with the atoms, i.e., by the atomic coefficient of weakening μ_a:

$$dI_x = -\mu_a I_x n dx, \tag{4.23}$$

where $n = \rho N/A$ is the number of atoms in the unit volume, ρ is the density of matter; $N = 6.023 \cdot 10^{23}$ mol^{-1} is the Avogadro number (the number of atoms in the gram atom), A is the atomic weight of matter.

Substituting the value n into (4.23) gives:

$$\frac{dI_x}{I_x} = -\mu_a \frac{\rho N}{A} dx.$$

After integration of this equation with respect x in range from 0 to x, we obtain:

$$\ln \frac{I_x}{I_0} = -\mu_a \frac{\rho N}{A} x,$$

or

$$I_x = I_0 \exp\left\{-\mu_a \frac{\rho N}{A} x\right\}. \tag{4.24}$$

The value $\mu_a \rho N/A$ in equation (4.24) is the linear coefficient of attenuation μ_x of x-radiation and has the dimension $[\mu_x] = cm^{-1}$, i.e., determines the attenuation of radiation in the unit length. Taking into account the notation, equation (4.24) is presented in the new form:

$$I_x = I_0 \exp\{-\mu_x x\}. \tag{4.25}$$

The linear coefficient of attenuation μ_x of x-radiation depends on the density of the absorbing matter and this is not always convenient. Therefore, the value μ_m, equal to $\mu_a N/A$, is usually used. In this case, from (4.24) we obtain:

$$I_x = I_0 \exp\{-\mu_m p_s\}, \tag{4.26}$$

where $p_s = \rho x$ is the surface density of the absorbing matter.

The value μ_m characterises the attenuation of radiation by the unit mass of the specimen and is referred to as the mass coefficient of attenuation. This coefficient has the dimension $[\mu_m] = cm^2 \cdot g^{-1}$. The ratio between the linear and mass coefficients of attenuation is determined by the expression:

$$\mu_x = \rho \mu_m.$$

The mass coefficient of attenuation of the x-radiation does not depend on density of matter, i.e., on its physical state. Therefore, the values of this coefficient may be tabulated for all elements and different wavelength ranges.

At present, there is a number of tables containing experimental and calculated values of the mass coefficients of attenuation. Detailed information on the experimental and calculated values of the coefficients of weakening, collected over many years by the National Institute of Standards and Technology of the USA, has been published in [166–168].

More accurate data on the coefficients of attenuation for relatively low-energy x-radiation may be found in [169, 170]. The coefficient of attenuation for a specific wavelength is determined by interpolation of the data presented in these tables. The coefficients of attenuation for characteristic x-radiation of all elements are also presented in a number of Russian studies (see, for example, [41, 171]).

In [33], the results of determination of the mass coefficients of attenuation are compared on the basis of tabulated data from 11 studies some of which are represented by the theoretically determined values of this coefficient (the data on the attenuation of radiation of petrogenous elements are compared). It has been shown that the accuracy of the calculated values is not inferior to the accuracy of the experimentally determined values, and this confirms the high level of the current theoretical considerations of the investigated processes.

A number of authors (see, for example, [172–175]) proposed the dependences approximating the experimental or calculated tabulated values of the coefficients of attenuation. They are based on the exponential equation:

$$\mu = C\lambda^n, \tag{4.27}$$

where the exponent n is tabulated, and value C is either constant for the given element or its dependence on the atomic number Z and λ is used.

In [176], the authors present in the polynomial dependences for calculating the attenuation of x-radiation of all elements in the energy range from 0.1 to 100 keV. The coefficients photoelectric absorption and also the integral coefficients of coherent and non-coherent scattering were calculated separately.

It is often necessary to calculate the attenuation of a substance of a complicated elemental composition. For example, we examine the solution of a problem of the determination of the coefficient of attenuation of radiation for some complex molecule. The molecular coefficient of attenuation is expressed by the atomic coefficient using the relationship:

$$\mu_{\text{mol}} = \sum_i (\mu_{\text{a}})_i n_i,$$

where the summation is carried out with respect to the atoms, forming the molecule, n_i is the number of these atoms in the molecule.

The molecular and atomic coefficients of attenuation are replaced by mass coefficients using the expressions $(\mu_m)_{\text{mol}}=\mu_{\text{mol}}\frac{N}{M}$ and $\mu_m=\mu_a\frac{N}{A}$. Consequently, we obtain:

$$(\mu_m)_{\text{mol}} = \sum_i \frac{n_i A_i}{M} \mu_{mi}.$$

However, the value $\frac{n_i A_i}{M}$ is the weight content C_i of the element i in the molecule. Therefore:

$$(\mu_m)_{\text{mol}} = \sum_i C_i \mu_{mi}.$$

The same method is used for deriving the equation for any homogeneous mixtures, alloys, solid and liquid solutions. Therefore, for homogeneous weakening media it is assumed that:

$$\mu_m = \sum_i C_i \mu_{mi}, \tag{4.28}$$

where C_i is the weight content of the i-th element in such a medium.

Chapter 5

Scattering of x-ray radiation

The scattering of x-ray radiation takes place on the atomic electrons of the irradiated material. A strong effect on the scattering process is exerted by the relationship of the scattering electrons with the nucleus and by their position in the atom at the moment of scattering.

The scattering of x-radiation, i.e., the variation of the direction of propagation of photons, may take place coherently, without any change of the energy (elastic interaction), and non-coherently (inelastic interaction) with the transfer of part of photon energy to one of the atomic electrons. Non-coherent scattering is accompanied by the formation of a flux of recoil electrons, whose kinetic energy depends on the angle of departure of the electrons. Initially, attention is given to the limiting case of the interaction of an x-ray photon with a free electron.

5.1. Scattering of x-ray radiation on free electrons

5.1.1. Coherent scattering

From the viewpoint of the classic theory, the resting electron is transferred to the state of forced vibrations under the effect of x-ray electromagnetic radiation . The electron becomes the source of radiation which also represents the scattered primary radiation. The frequency of the forced vibrations of the electron is equal to the frequency of primary radiation. In other words, according to the classic theory, the scatter of x-ray radiation is coherent and has the same wavelength as primary radiation.

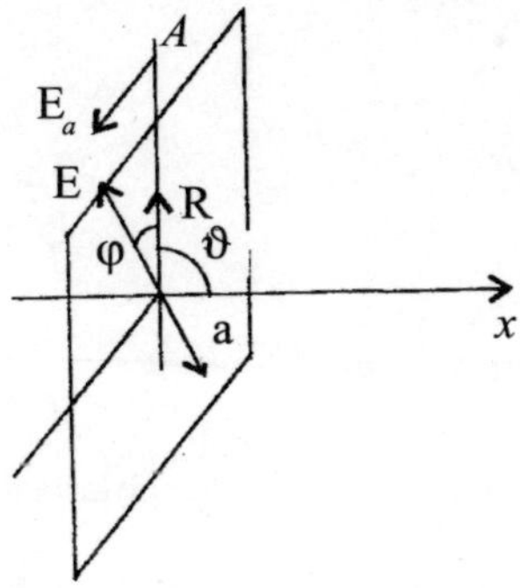

Fig. 5.1. Diagram of scattering of x-ray radiation in the plane, normal to the direction of the primary beam. The specific feature of coherent scattered radiation is its polarisation.

It is assumed that x-radiation propagates along the axis x and is scattered in the direction of the vector **R**, normal to the direction of the primary beam (Fig. 5.1). The electrical vector **E** is normal to the direction of propagation of radiation and, therefore, in the investigated case is positioned with the vector **R** in the same plane, normal to the primary beam. The direction of acceleration vector **a** is opposite of the direction of vector **E** and acceleration vector **a** is also situated in the same plane.

The strength of the electrical field $\mathbf{E}_A$ (determined by the forced vibrations of the electron) at the examination point A is determined by the equation:

$$\mathbf{E}_A = \frac{e\mathbf{a}}{c^2 R}\sin\varphi,$$

where angle φ is the angle between the vector **R** and the acceleration vector **a** of the scattering electron.

Acceleration **a** of the electron depends on the strength E of the primary electrical field, since

$$\mathbf{a} = \frac{e\mathbf{E}}{m}.$$

Therefore, we obtain the following equation for the strength of the field $\mathbf{E}_A$:

$$E_A = \frac{e^2}{mc^2}\frac{\mathbf{E}}{R}\sin\varphi. \qquad (5.1)$$

The value of angle φ is determined for the case of polarised primary radiation. The angle has not been determined for the non-polarised primary radiation. Regardless of this, the vector $\mathbf{E}_A$ is always normal to the vector **R** and its orientation does not depend on the direction of the vector **E**. In other words, the non-polarised primary x-ray radiation, scattered in the direction normal to the direction of the primary beam, is completely polarised.

The polarisation of coherently scattered x-ray radiation was discovered

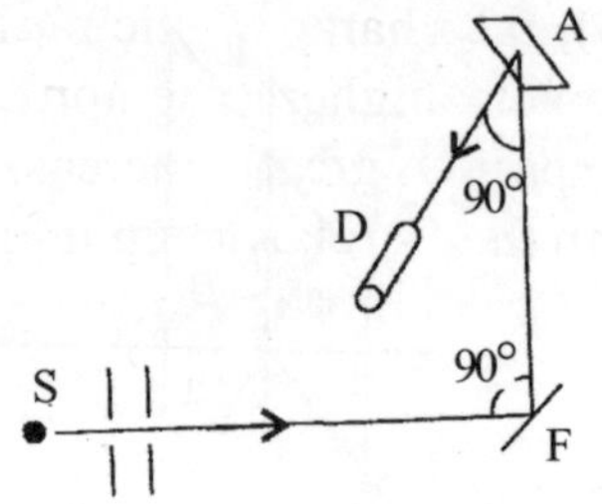

Fig. 5.2. Diagram of experiments with the polarisation of coherently scattered x-ray radiation.

experimentally for the first time by Barclay in 1905. The experiments carried out by Barclay confirmed the hypothesis that the x-ray radiation is electromagnetic, i.e., it is of the wave nature. The diagram of equipment used for these experiments is shown in Fig. 5.2. Here S is the source of primary non-polarised x-radiation, F is the first scatterer, A is the second scatterer, D is the radiation detector. If the scattering matter has a small atomic number, the electrons of the outer shells of the matter may be regarded in the first approximation as free.

In these conditions, the angle φ between the acceleration vector of the electrons of the second scatterer and the direction from A to D is equal to zero. Correspondingly, the intensity of scattered radiation is also equal to zero.

In the experiments, carried out by Barclay, the intensity of radiation, recorded by the detector D, was 3.3 times smaller than the intensity measured by the detector under the same angle of 90° in the direction FA, but with the detector positioned in the plane including the points S, F and A. The increase of the polarisation capacity of the Barclay system assumes more stringent requirements on the collimation of beams and suppression of the process of multiple scattering by optimisation of the thickness and composition of scattering media [177].

The experimental setup, shown in Fig. 5.2, may be used for increasing the contrast of the fluorescent signal by reducing the intensity of background radiation, determined by coherent scattering. In fact, the fluorescence of the object, placed at point A, is excited by polarised primary radiation. The intensity of fluorescence does not depend on the degree of polarisation of exciting radiation, and the scattered radiation in the direction AD attenuates.

Difficulties in the practical realisation of this approach arise because of the low intensity of scattered polarised radiation. If the polariser F is represented by a single crystal with the plane spacing being such that the scattering angle is 90° and, at the same time, corresponds to the direction of 'reflection' of characteristic radiation of the source S according to Wulf–

Bragg law (section 5.5), the characteristic radiation will be polarised and its intensity is considerably higher. The application of this radiation for exciting fluorescence at point A greatly increases the contrast of fluorescent radiation. The interest in the Barclay setup in this combination with Bragg reflection increased considerably when the effect of complete external reflection was used in x-ray fluorescence analysis [177] when the take-off angle of fluorescent radiation of the specimen is practically equal to 90° in relation to the direction of the strictly collimated primary beam.

Another possibility of increasing the intensity of polarised radiation at point A is the application of the Barclay polariser setup, shown in Fig. 5.3.

In this setup the polariser is represented by the cylindrical surface of a material with a high scattering capacity (for example, graphite). A source of the diverging beam of non-polarised x-radiation is placed on the surface, and the fluorescent object is located on the diametrically opposite side of the surface. All the scattering angles in the polariser are equal to 90°. This corresponds to the condition of complete polarisation of coherent radiation. The direct incidence of primary radiation from the source S on the fluorescent object is prevented by placing the screen E. If the fluorescence of the object A is recorded in the direction normal to the plane of the figure, the background, determined by the coherent scattering of polarised radiation, will be minimal. The Barclay polariser makes it possible to increase the intensity of polarised radiation by more than an order of magnitude in comparison with the case shown in Fig. 5.2.

The classic theory of the formation of the intensity of scattered x-ray radiation was proposed by Thomson [178].

Since the intensity of radiation is proportional to the square of the strength of the electromagnetic field, equation (5.1) gives the following formula for the ratio of the intensities of scattered and primary non-polarised radiation:

$$\frac{I_\varphi}{I} = \frac{E_a^2}{E^2} = \frac{e^4}{m^2 c^4 R^2} \sin^2 \varphi, \tag{5.2}$$

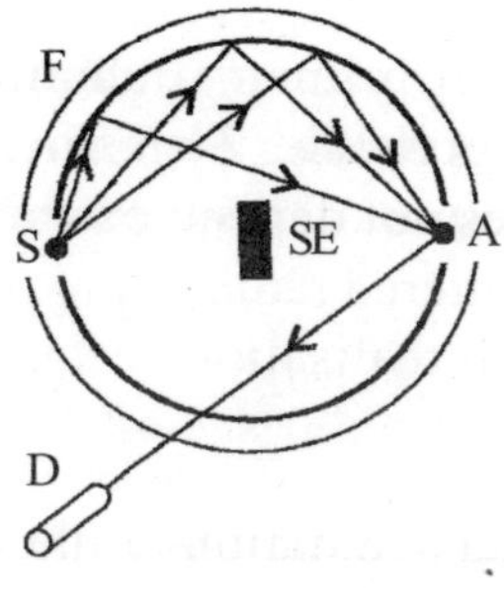

Fig. 5.3. The Barclay polariser.

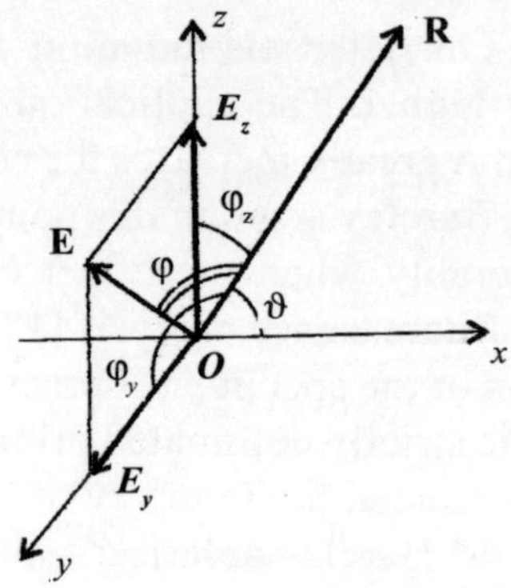

Fig. 5.4. Diagram of determination of the intensity of coherently scattered x-ray radiation.

where I_φ is the intensity of scattered radiation.

For the non-polarised primary radiation and the arbitrary position of the observation point, the formation of the intensity of scattered x-radiation is determined by the setup shown in Fig. 5.4. It is assumed that x-radiation, propagating in the direction of the axis x, is scattered on a free electron, situated at point O. The scattered radiation is detected at point A situated at a distance **R** from the scattering electron. The scattering angle, i.e., the angle between the axis x and vector **R**, is denoted by ϑ. In the plane of the vector **R** and the axis x, we draw through the point O the axis z normal to the axis x. At the same point O, we place the perpendicular y to the plane, containing the axis x and z.

Vector **E** of the strength of the field is normal to the direction of propagation of radiation and, consequently, is located in the plane yz. The projections of this vector on the axes y and z are E_y and E_z, respectively. Here

$$E^2 = E_y^2 + E_z^2.$$

The angles between the direction of scattering **R** and the axes y and z are equal to φ_y and φ_z, respectively.

In the case of non-polarised primary radiation, the direction of the electrical vector **E** is equally probable. Therefore, the following equation may be proposed for the projections of this vector averaged out with respect to all directions:

$$E_y^2 = E_z^2.$$

Consequently, $E^2 = 2E_y^2 = 2E_z^2$.

Thus, the intensity of primary radiation at the scattering point is equal to:

$$I = 2I_y = 2I_z. \tag{5.3}$$

Both components $I_{\varphi y}$ and $I_{\varphi z}$ of scattered radiation I_φ, detected at point A, may be described by the following equation on the basis of (5.2):

$$I_{\varphi y} = I_y \frac{e^4}{m^2c^4R^2}\sin^2\varphi_y;$$

$$I_{\varphi z} = I_z \frac{e^4}{m^2c^4R^2}\sin^2\varphi_z.$$

Consequently, taking into account that $\varphi_y=\pi/2$ and $\varphi_z=\pi/2-\vartheta$, and taking into account equation (5.3), we obtain:

$$I_{\varphi y} = \frac{1}{2}I\frac{e^4}{m^2c^4R^2};$$

$$I_{\varphi z} = \frac{1}{2}I\frac{e^4}{m^2c^4R^2}\cos^2\vartheta.$$

Finally, for the intensity of scattered radiation we have:

$$I_T = I_{\varphi y} + I_{\varphi z} = I\frac{r_0^2}{R^2}\frac{1+\cos^2\vartheta}{2}, \tag{5.4}$$

where $r_0 = \dfrac{e^2}{mc^2}$ is the classic radius of the electron or its Thomson length (r_0^2 =7.94·10⁻²⁶ cm²).

The multiplier (1+cos²ϑ)/2 in equation (5.4) is the polarisation factor and characterises the effect of polarisation on the intensity of coherently scattered x-ray radiation. As mentioned previously, in the direction $\vartheta = \pi/2$ the radiation is completely polarised. There is no polarisation in the direction $\vartheta = 0$ and $\vartheta = \pi$.

The total electronic coefficient of coherent scattering of x-ray radiation may be determined by integrating equation (5.4) with respect to all possible directions of the radius-vector **R**. In polar coordinates, integration is carried out with respect to the angle ϑ in the range from 0 to π for the elementary sections of the sphere $2\pi rRd\vartheta = 2\pi R^2 \sin\vartheta d\vartheta$. Consequently, we obtain:

$$\frac{W_T}{t} = I\pi r_0^2\int_0^\pi (1+\cos^2\vartheta)\sin\vartheta d\vartheta = \frac{8}{3}I\pi r_0^2,$$

where

$$\sigma_e^T = \frac{W_T / t}{I} = \frac{8}{3}\pi r_0^2. \tag{5.5}$$

Intensity I_T in equation (5.4) is distributed over the area dA of the surface of the front of the diverging wave within the limits of the small solid angle $d\Omega$ at distance R from the scattering point. Therefore, the differential electronic coefficient of Thomson scattering (the energy, scattered by a single electron in the direction ϑ in the range of the angle $d\Omega$), is determined as follows:

$$\frac{d\sigma_e^T}{d\Omega} = \frac{I_T}{I}\frac{dA}{d\Omega} = \frac{r_0}{2}(1+\cos^2\vartheta). \tag{5.6}$$

Coherent scattering is sometimes referred to as Rayleigh scattering, as a result of the investigations carried out by Lord Rayleigh (pseudonym Strutt) at the end of the 19th century into the processes of scattering of polarisation of visible light by gas molecules [170, 180].

5.1.2. Non-coherent scattering

The quantum theory of non-coherent scattering of x-ray radiation on free electrons was proposed by Compton [2, 181]. This scattering process is shown in Fig. 5.5.

The photon, indicent at point O on the resting electron, has the energy $E = h\nu$ and the momentum $p = h\nu/c$. After scattering under the angle ϑ, the photon receives the energy $E' = h\nu'$ and the momentum $p = h\nu'/c$.

In turn, the electron acquires the energy E_{kin} and the momentum $p_e = m\upsilon$.

On the basis of the law of conservation of energy:

$$E_{\text{kin}} = E - E', \tag{5.7}$$

where E_{kin} is the difference between the total energy and the rest energy of the electron:

$$E_{\text{kin}} = mc^2 - m_0c^2 = m_0c^2\left(\frac{m}{m_0} - 1\right) = m_0c^2\left(\frac{1}{\sqrt{1-\beta^2}} - 1\right).$$

Substituting the resultant expression into (5.7), we obtain:

$$E - E' = m_0c^2\left(\frac{1}{\sqrt{1-\beta^2}} - 1\right).$$

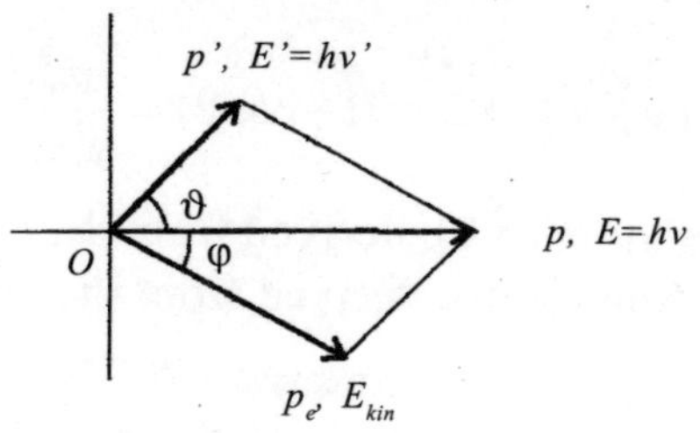

Fig. 5.5. Non-coherent scattering of an x-ray photon on a free electron.

Otherwise $h\nu - h\nu' = \frac{m_0c^2}{\sqrt{1-\beta^2}} - m_0c^2$. Consequently $(\nu - \nu') + \frac{m_0c^2}{h} = \frac{m_0c^2}{h\sqrt{1-\beta^2}}$.

Squaring both parts of the equality, we obtain:

$$(\nu - \nu')^2 + 2\frac{m_0c^2}{h}(\nu - \nu') + \frac{m_0^2c^4}{h^2} = \frac{m_0^2c^4}{h^2(1-\beta^2)}.$$

or

$$(\nu - \nu')^2 + 2\frac{m_0c^2}{h}(\nu - \nu') = \frac{m_0^2c^4\beta^2}{h^2(1-\beta^2)}. \quad (5.8)$$

On the basis of the law of conservation of the pulse:

$$\mathbf{p} - \mathbf{p}' = \mathbf{p}_e.$$

We determine the scalar product of these vectors:

$$((\mathbf{p} - \mathbf{p}')(\mathbf{p} - \mathbf{p}')) = (\mathbf{p}_e)^2,$$

and therefore

$$p^2 + (p')^2 - 2pp'\cos\vartheta = p_e^2. \quad (5.9)$$

Taking into account the previously discussed values p, p' and p_e, we obtain:

$$\left(\frac{h\nu}{c}\right)^2 + \left(\frac{h\nu'}{c}\right)^2 - 2\frac{h^2}{c^2}\nu\nu'\cos\vartheta = m^2\upsilon^2 = \frac{m_0^2c^2\beta^2}{(1-\beta^2)},$$

from which

$$\nu^2 + \nu'^2 - 2\nu\nu'\cos\vartheta = \frac{m_0^2c^4\beta^2}{h^2(1-\beta^2)},$$

or

$$(\nu-\nu')^2+2\nu\nu'(1-\cos\vartheta)=\frac{m_0^2c^4\beta^2}{h^2(1-\beta^2)}. \qquad (5.10)$$

Comparison of the equation (5.8), derived from the law of conservation of energy, with the equation (5.10), derived from the law of conservation of the momentum, gives:

$$2\nu\nu'(1-\cos\vartheta)=2\frac{m_0c^2}{h}(\nu-\nu'),$$

from which

$$\frac{(\nu-\nu')}{\nu\nu'}=\frac{h}{m_0c^2}(1-\cos\vartheta). \qquad (5.11)$$

Since $\nu = E/h$, equation (5.11) gives the relationship between the energies of primary and scattered photons:

$$E'=\frac{E}{1+\frac{E}{m_0c^2}(1-\cos\vartheta)}. \qquad (5.12)$$

We transfer to the examination of the wavelength of scattered radiation, using the dependence $\nu = c/\lambda$. From (5.11):

$$(\lambda'-\lambda)=\frac{h}{m_0c}(1-\cos\vartheta),$$

or

$$\Delta\lambda=\lambda_C(1-\cos\vartheta), \qquad (5.13)$$

where $\lambda_C=\frac{h}{m_0c}=0.024$ Å is the Compton electron length [129].

This equation shows that the variation of the wavelength at scattering of the photon on the free electron does not depend on the wavelength λ of scattering radiation, and is determined only by the angle ϑ under which this scattering took place.

At $\vartheta=\pi/2$, $\Delta\lambda_\perp=\lambda_C=0.024$ Å.

The variation of the wavelength by 0.024 Å is quite appreciable for shortwave radiation. For $\lambda = 0.024$ Å the scattering under the angle of 90° results in doubling of the wavelength, and in scattering under the angle of 180° the wavelength increases three times. For the long-wave range, this variation is not so large. For example, if $\lambda = 2.4$ Å, then at $\vartheta = 90°$, the wavelength changes by only 1%.

It should be remembered that in contrast to the variation of the wavelength, the variation of the photon energy depends on the energy of the photon itself.

In fact, equation (5.12) shows that:

$$E - E' = \Delta E = \frac{\frac{E^2}{m_0c^2}(1-\cos\vartheta)}{1+\frac{E}{m_0c^2}(1-\cos\vartheta)}. \tag{5.14}$$

In equation (5.14), m_0c^2 is the rest energy of the electron, equal to 512 keV. Therefore, at a low energy of the scattering photons, the second term in the denominator may be ignored. This gives:

$$\Delta E \approx \frac{E^2}{m_0c^2}(1-\cos\vartheta). \tag{5.15}$$

This means that in the case of a low energy of the x-ray photons, the variation of the energy in the Compton scattering of a free electron is approximately proportional to the square of the photon energy.

5.1.3. Intensity of x-ray radiation, scattered on free electrons

The Compton theory does not make it possible to estimate the intensity of scattered radiation and its spatial distribution. These characteristics for the scattering of the x-ray photon on the free electron are determined by the Klein–Nishina–Tamm theory [182], developed on the basis of quantum electrodynamics. This theory shows that the intensity of scattered radiation may be determined from the equation:

$$I^{\mathrm{KNT}} = I\frac{r_0^2}{2R^2}\left(\frac{E'}{E}\right)^2\left(\frac{E}{E'}+\frac{E'}{E}-\sin^2\vartheta\right), \tag{5.16}$$

where r_0 is the classic radius of the electron (see the previous section), E and E' is the energy of the x-ray photon prior to and after scattering, respectively, ϑ is the scattering angle, R is the distance from the area of scattering to the observation point.

The dependence of the intensity of scattered radiation on the scattering angle ϑ is complicated because in equation (5.16) we must take into account the effect of the scattering angle on the ratio of the values E and E', given by equation (5.12). The polar diagram of the spatial distribution of scattered x-ray radiation is shown in Fig. 5.6.

The graph shows that at high photon energies, scattering of photons takes place in the direction close to 0° (the direction of the primary photon beam).

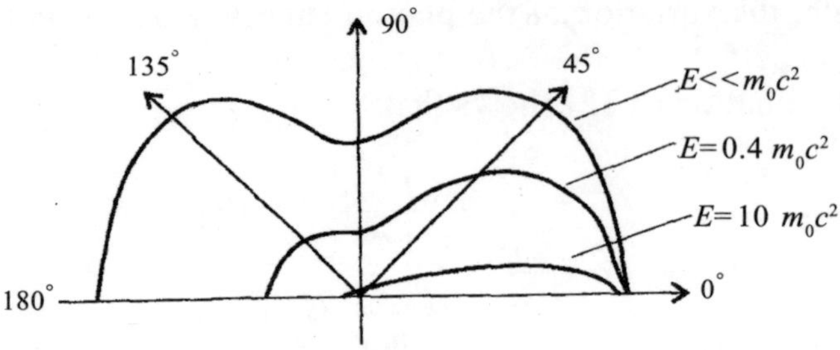

Fig. 5.6. Dependence of the spatial distribution of scattered x-ray radiation on the energy of its photons.

At a comparatively low photon energy, when $\Delta E \to 0$ (see the equation (5.15)) and, correspondingly, $E \to E'$, equation (5.16) gives the previously examined equations for the intensity of coherent (Thomson) scattering of x-ray radiation:

$$I^{\mathrm{T}} = I\frac{r_0^2}{R^2}\frac{2-\sin^2\vartheta}{2} = I\frac{r_0^2}{R^2}\frac{1+\cos^2\vartheta}{2} \tag{5.17}$$

Thus, at low photon energy, the distribution of scattered radiation becomes symmetric in relation to the plane, normal to the direction of propagation of the beam. The intensity at $\vartheta = \pi/2$ is half the maximum value of intensity observed in the direction $\vartheta = 0°$ and $\vartheta = \pi$ (see the case $E \ll m_0c^2$ in Fig. 5.6).

The expression for the differential electronic coefficient of scattering, for the unit solid angle dΩ, is derived from the main dependence (5.16) by transformations identical with those carried out in the case of coherent scattering. This gives:

$$\frac{d\sigma_e^{\mathrm{KNT}}}{d\Omega} = \frac{r_0^2}{2}\left(\frac{E'}{E}\right)^2\left(\frac{E}{E'}+\frac{E'}{E}-\sin^2\vartheta\right), \tag{5.18}$$

where $d\Omega$ is the elementary solid angle in the direction ϑ.

For non-coherent scattering, the value of the total electronic coefficient of scattering σ_e^{KNT} may be determined by the numerical integration of equation (5.18) with respect to all angles ϑ for some energy E' of scattered radiation. The dependence σ_e^{KNT} on the energy of the scattering photons is shown in Fig. 5.7 (in the units of the integral electronic coefficient σ_e^{T} of coherent scattering).

Figure 5.7 shows that the total coefficient of Compton scattering decreases with increasing photon energy and, consequently, the intensity of scattering decreases (this corresponds to the data in Fig. 5.6).

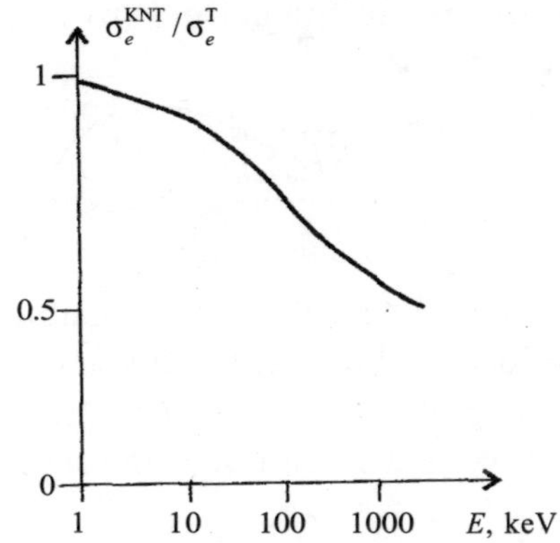

Fig. 5.7. Dependence of the electronic coefficient of non-coherent scattering on photon energy (in units σ_e^T).

5.2. The scattering of x-ray radiation on atoms

In the real conditions, the x-ray photon interacts with the bonded electrons of the atoms. This may be accompanied by the interaction of the photon both with the individual electron of the atom (Compton effect) and with the atom as a whole. Since the mass of the atom is large in comparison with the mass of the electron, in the latter case the recoil effect (the magnitude of the momentum) for the atom is negligibly small and the scattering of the photon takes place almost coherently, without any change of the energy.

In the examination of the process of scattering of x-ray radiation by the one-electron atom [2, 7], it is assumed that the electron may be situated with certain probability at any point in the vicinity of the atomic nucleus. On the basis of the condition of symmetry of the distribution of this probability in relation to the centre of the sphere (or the position of the atomic nucleus), the probability of the electron being in some spherical layer of the atom shell is determined by the quantity $P(r)$, where r is the distance of the electron from the atomic nucleus. In the classic approach, the intensity of the radiation, scattered by this electron, at some observation point, far away from the atom, is determined by the Umov–Pointing equation

$$I_a = \frac{c}{4\pi}\left|\overline{(E_\vartheta)_\varepsilon}\right|^2, \tag{5.19}$$

where $\overline{(E_\vartheta)_\varepsilon}$ is the strength of the electrical field, averaged out with respect to time, at the observation point under the angle ϑ to the initial direction for the radiation scattered by the electron with the charge ε 'spread' over the atomic volume.

It is rational to examine the factor of the electronic structure, f, determined as the ratio of the intensity I_a to the intensity of coherent scattering by the free electron:

$$f^2 = \frac{I_a}{I_T} = \frac{\left|\overline{(E_\vartheta)_\varepsilon}\right|^2}{\left|E_\vartheta\right|^2}. \tag{5.20}$$

This factor is calculated by integrating the probability $P(r)$ of the electron being situated in the local volume at the distance r from the nucleus. In integration, various dependences of the values $\overline{(E_\vartheta)_\varepsilon}$ and E_ϑ on the scattering angle ϑ and the wavelength λ of the scattered radiation are also taken into account:

$$f = \int_0^\infty P(r)\frac{\sin \varsigma r}{\varsigma r}\,dr, \tag{5.21}$$

where

$$\varsigma = \frac{4\pi}{\lambda}\sin\frac{\vartheta}{2}. \tag{5.22}$$

We now examine a many-electron atom with the number of electrons equal to the charge Z of the atomic nucleus. In the quantum mechanics approach, the probability of the electron being situated in some volume in the vicinity of the atomic nucleus is determined by the square of the wave function of the electron ψ_n which may be calculated using the method of the self-consistent Hartree–Fock field or by the Thomas–Fermi approximate methods.

The radiation, scattered by the individual electrons, propagates in different directions with the different phases. Summing up the amplitude of the oscillations of the electrical vector of radiation of the individual electrons and taking their interaction into account, for the ratio of the intensity, scattered by all electrons of the atom, to the intensity I_e, scattered by the free electron, we obtain the following expression [2]:

$$\frac{\overline{I_a}}{I_e} = \sum_{n=1}^{Z}(1-|f_n|^2) + \left|\sum_{n=1}^{Z} f_n\right|^2, \tag{5.23}$$

where f_n is the factor of the electronic structure for the n-th atomic electron, determined by the equation (5.21).

The application of the principle of correspondence of the classic and quantum mechanics approaches shows that the first sum in the equation (5.23) determines the non-coherent part of the scattered radiation, and the second sum its coherent. This means that:

$$\frac{I^{\mathrm{c}}}{I^{\mathrm{T}}} = \left| \sum_{n=1}^{Z} f_n \right|^2 \tag{5.24}$$

and

$$\frac{I^{\mathrm{nc}}}{I^{\mathrm{KNT}}} = \sum_{n=1}^{Z} (1 - |f_n|^2). \tag{5.25}$$

We cxamine the coherent and non-coherent scattering separately.

5.2.1. Coherent scattering by atoms

The intensity of coherently scattered radiation, determined by the expression (5.24), contains the sum $\sum_{n=1}^{Z} f_n$, referred to as the atomic form-factor F.

Equation (5.24) shows that:

$$I^{\mathrm{c}} = I^{\mathrm{T}} F^2, \tag{5.26}$$

where the intensity I^{T} of the coherent scattering on the free electron is determined by the equation (5.17).

Thus, the intensity of coherent scattering for the atom is obtained by multiplying the intensity of coherent scattering for the free electron by the square of the atomic form-factor F.

The identical dependence obtained for the atomic differential coefficient of coherent scattering:

$$\frac{d\sigma_a^{\mathrm{T}}}{d\Omega} = \frac{d\sigma_e^{\mathrm{T}}}{d\Omega} F^2. \tag{5.27}$$

The accurate calculation of factor F is performed on the basis of the quantum mechanics considerations by the method of the self-consistent Hartree–Fock field.

The results of these calculations are presented in the literature in the form of tables as a function of two variables:

$$F = F(x, Z),$$

where the variable x is determined from (5.22) as:

$$x = \frac{\zeta}{4\pi} = \frac{1}{\lambda} \sin\frac{\vartheta}{2}. \tag{5.28}$$

The compiled tabulated values of the atomic form factor in the non-relativistic variant have been published, in particular, in [183]. In compiling the values, the data in [184–186] were used. The relativistic values of the form factor are presented in [187], where the materials of the investigations [188–190] were used.

Marenkov [191] used the non-relativistic form-factors $F(x, Z)$ from [183] for the determination of the integral coefficients σ_a of the atomic cross-section of coherent scattering. The dependence of σ_a on the energy E of the scattering photons should be calculated using the following approximating equation:

$$\sigma_a = (1+\chi_4 E)(\chi_0 + \chi_1 E + \chi_2 E^2 + \chi_3 E^3)^{-1}, \tag{5.29}$$

where the values of the coefficients χ_i are tabulated for every element.

The error of approximation in relation to the data published in [183] in the energy range 0.1–150 keV does not exceed in the majority of cases the value 2%.

In addition to the previously examined variants, the atomic form factor may also be calculated by the approximate Thomas–Fermi method. In this case, the atomic form factor is represented by the function of a single variable:

$$g = \left(\frac{5.91}{\lambda}\right)\sin\left(\frac{\vartheta}{2}\right)Z^{-1/3}. \tag{5.30}$$

It may easily be seen that this variable copies the dependence of the quantity ζ, given by the equation (5.22), on the scattering angle ϑ and the wavelength λ of scattered radiation, supplementing it by taking into account the atomic number Z of the scattering element.

The difference between the dependence of the value of the form factor F on Z according to Hartree–Fock and the approximate calculations of this dependence using the Thomas–Fermi method is of the systematic nature. This is explained by the fact that the structure of the atom is not taken into account sufficiently using the function $g{\sim}Z^{-\frac{1}{3}}$. Better agreement with the calculations using the Hartree–Fock method may be achieved by selecting [192] the following approximating equation:

$$F \approx Z(e^{-g} + 0.12). \tag{5.31}$$

The graphical dependence $F = f(g)$ is shown in Fig. 5.8.

From the viewpoint of the classic theory, the combination of the natural oscillations of the electrons and the forced oscillations of the electrons changes the phase of the resultant wave. The degree of distortion of the resultant wave increases with a decrease of the difference between the frequency of the intrinsic and forced oscillations. When these frequencies coincide, the forced oscillations of the electrons have the phase opposite to the phase of the intrinsic oscillations, because in the case of a negative charge of the electron, the direction of movement of the electron is opposite to the direction of the electrical vector of the external electromagnetic field. Thus, in the region of the absorption edge of the scattering element, there is anomalous scattering, and to take this scattering into account, it is

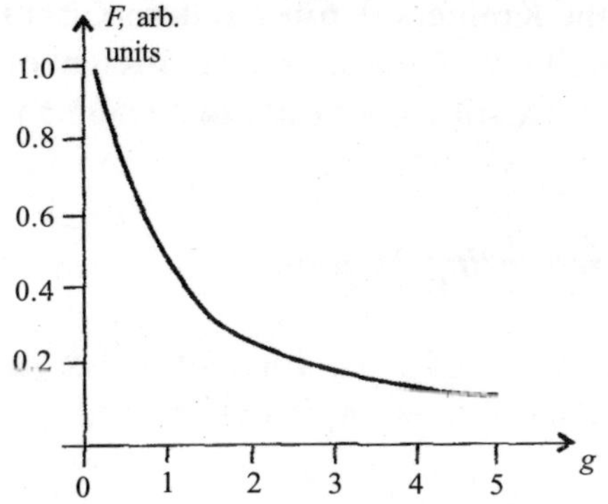

Fig. 5.8. Dependence of the normalised atomic factor F on argument g.

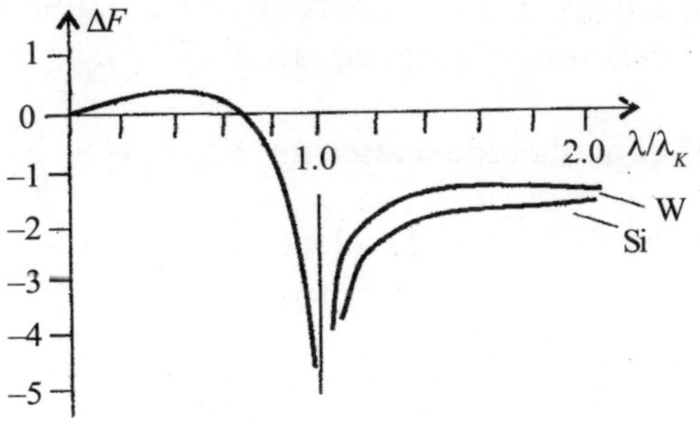

Fig. 5.9. Dependence of the correction ΔF on anomalous scattering on the wavelength in the vicinity of the K-edge of adsorption of the elements W and Si.

necessary to introduce correction ΔF into the equation (5.26). The dependence of this correction on the wavelength λ in the vicinity of the K-edge of adsorption is shown in Fig. 5.9.

Figure 5.9 shows that at $\lambda < 0.9\lambda_K$, the anomalous scattering of x-ray radiation may be ignored.

At $0.9\ \lambda_K < \lambda < 1.1\lambda_K$ it is necessary to introduce a correction for anomalous scattering:

$$I^c = I^T (F + \Delta F)^2. \tag{5.32}$$

At $1.1\lambda_K < \lambda$, the anomalous scattering is taking into account approximately by the decrease of the atomic form-factor by 2:

$$I^c = I^T (F - 2)^2. \tag{5.33}$$

The quantum-mechanical calculation of the cross-section of coherent scattering σ_a^T in the anomalous region in the vicinity of the adsorption edges requires ignoring the considerations of the self-consistent fields of the Hartree–Fock theory and is based on the application of the relativistic S-matrix of the second order [193–195] (see the following section).

The application of the Kramers–Kronig relation, linking the correction for the anomalous scattering with the atomic coefficient of absorption, gives [129] the value of the correction ΔF equal to −1.565 for the wavelengths $\lambda \gg \lambda_K$.

5.2.2. Non-coherent scattering by atoms

As indicated by equation (5.25), the intensity of x-ray radiation, non-coherently scattered by the free atom, is determined by the dependence:

$$I^{nc} = I^{KNT}\left(\sum_{n=1}^{Z}(1-|f_n|^2)\right) = I^{KNT}Z\left(1-\frac{1}{Z}\sum_{n=1}^{Z}|f_n|^2\right) = I^{KNT}H, \qquad (5.34)$$

where I^{KNT} is the intensity of non-coherent scattering on the free electron, given by the Klein–Nishima–Tamm equation (5.16).

The function of noncoherent scattering $H = Z\left(1-\frac{1}{Z}\sum_{n=1}^{Z}|f_n|^2\right)$ in equation (5.34) depends on the factors of the electronic structure f_n, determined by the equation (5.21). Therefore, $H(Z, x)$, like the form-factor $F(Z, x)$ in coherent scattering, is a function of two variables Z and $x = \frac{1}{\lambda}\sin\frac{\vartheta}{2}$, the latter of which is determined by the wavelength λ of the scattered radiation and by the scattering angle ϑ.

The differential atomic coefficient of scattering $d\sigma_a^{nc}/d\Omega$ is determined from the electronic coefficient of non-coherent scattering $d\sigma_e^{KNT}/d\Omega$, determined by the equation (5.18), by multiplying this equation by the function H of non-coherent scattering:

$$\frac{d\sigma_a^{nc}}{d\Omega} = \frac{d\sigma_e^{KNT}}{d\Omega}H. \qquad (5.35)$$

The results of non-relativistic theoretical calculations using the Hartree–Fock method, and also the experimental determination of the non-coherent scattering function H and of the atomic cross-section of non-coherent scattering σ_a^{nc} are presented in the form of tables [183], based on the data compiled from [184, 185, 196, 197].

The results of relativistic calculations of the atomic cross-section of non-coherent scattering [198, 199] by the Dirac–Hartree–Fock method also assume that the function of non-coherent scattering H depends on two parameters Z and x.

The application of the approximate Thomas–Fermi method enables the value of H/Z to be calculated as a function of only one variable:

$$\upsilon = \frac{2.21}{\lambda} \sin\left(\frac{\vartheta}{2}\right) Z^{-2/3}.$$

The error of approximate calculations of the values of H/Z using the parameter υ in the range $0.1 < \upsilon < 0.3$ in relation to the calculations using the Hartree–Fock method is greater than when using the approximation equation, presented in [192]:

$$\frac{H}{Z} = 1 - e^{-5\upsilon}.$$

Consequently, the following equation may be used for the approximate calculations of the noncoherently scattering intensity of x-ray radiation at $\upsilon > 0.1$:

$$I^{\text{nc}} = I^{\text{KNT}} Z(1 - e^{-5\upsilon}), \tag{5.36}$$

and for the calculation of the atomic coefficients of non-coherent scattering:

$$\frac{d\sigma_a^{\text{nc}}}{d\Omega} = \frac{d\sigma_e^{\text{KNT}}}{d\Omega} Z(1 - e^{-5\upsilon}). \tag{5.37}$$

More accurate calculations of the intensity of non-coherent scattering are linked with the application of the collisions matrix (S-matrix) introduced into the theory of quantum scattering independently by Wheeler [200] and Heisenberg [201, 202]. Processing and application of the S-matrices have been examined in a number of investigations (for example, [203]).

The concept of the S-matrix results from the examination of the Hamiltonian of the interaction of the quantum-mechanics system, consisting of interacting objects (the atom and the electromagnetic field, the atom and the incident proton, etc). The initial state of the system is expressed by the Slater determinant, having the form of a linear combination of the electronic eigenfunctions ψ_n. The final state of the system (after interaction) may be represented by a second linear combination of the same set of eigenfunctions. If the wave function on the initial state is denoted by $\Psi(t = -\infty)$, and the wave function of the final state by $\Psi(t = +\infty)$, they are expressed by the eigenfunctions of the electrons as:

$$\Psi(t = -\infty) = \sum_n a_n \psi_n$$

and

$$\Psi(t = +\infty) = \sum_m b_m \psi_m.$$

In this case, the coefficients b_m of the eigenfunctions of the final state is expressed by the coefficients a_n of the eigenfunctions of the initial state:

$$b_m = \sum_n S_{mn} a_n \tag{5.38}$$

using elements S_{mn}, which also form the so-called S-matrix. Thus, the S-matrix determines the process of interaction of the photon with the atom without any simplifying assumption on the spherical symmetry of the effective potential of the atomic electrons, which forms the basis of the Hartree–Fock method.

The practical application of the formalism of the S-matrices proved to be possible only when increasing the efficiency and the rate of computing operations, associated with the application of computers.

The development of the concept of the S-matrices in the direction of producing the S-matrices of the second order for the relativistic calculations of the cross-section of coherent scattering of the photons on the bonded electrons with the independent approximation of the particles has been carried out in [204–208] and summarised in a review in [209]. These calculations provide more accurate results in comparison with the non-relativistic data, tabulated in [183].

At the same time, the application of the collision matrix results in certain difficulties in the representation of the results in the form suitable for practical application, because the two-dimensional set of the function $H(Z, x)$ of the non-coherent scattering is replaced by a three-dimensional function $H(\lambda, \vartheta, Z)$ which may be realistically tabulated only in the form of the data, calculated in the computer.

The characteristic radiation, non-coherently scattered on the electrons of the atoms, is observed in the form of a line of greatly increased thickness. This broadening is not in agreement with Compton' s theory which does not take into account the relationship of the scattering electrons with the nucleus and the movement of the electrons on the orbits. The expression for increasing the width of the non-coherently scattered line, determined by these reasons, was derived by Wentzel [210, 211] from the theoretical assumptions of wave mechanics:

$$(\delta E)^2 \approx E_b \Delta E, \tag{5.39}$$

where δE is the width of the non-coherent line at half its maximum, E_b is the binding energy of the electron on which scattering takes place, ΔE is the variation of the photon energy in Compton scattering.

Approximately the same result was obtained by DuMond [212] in the examination of the Doppler effect in the projection of the velocity of the electrons on the direction of the x-ray radiation beam.

Equation (5.39) was used in [31] for evaluating the expansion of MoKα-line (E = 17.5 keV), non-coherently scattered on carbon under the angle ϑ = 90°. According to equation (5.10), the change of the energy of incident radiation is $\Delta E = 560$ eV. The binding energy of the K-electrons of carbon is $E_b = E_K = 350$ eV. The calculated values of expansion δE is 450 eV. This means that the expansion of the non-coherently scattered line is quite considerable and comparable in this case with the Compton shift.

Early equations for the broadening of the non-coherent scattering line, similar to (5.39), and other equations, proposed in a number of investigations [6, 212, 213], describe very approximately the form of these lines referred to as the Compton profile. The previously examined quantum mechanical models proposed at a later date (the Dirac–Hartree–Fock approximation and the theory of the S-matrices) indicate that it is possible to calculate the Compton profile for the given electronic structure, determined not only by the structure of the atom but also by the special features of interatomic interactions. Reaching agreement between the calculated and experimental Compton profile, information is obtained on the structure of the electron shells, chemical bonds and other molecular parameters of the investigated compound. The rapid determination of the experimental Compton profiles high-resolution spectrometers assumes the application of high intensity sources of x-ray synchrotron radiation. Some of the results of the investigations of the atomic and molecular structure on the basis of the profile of Compton scattering have been published in [214–217].

5.2.3. The ratio of the intensities of coherent and non-coherent scattering on the atoms

The previously mentioned results of investigations indicate that it is possible to calculate in each specific case the intensity or scattering coefficients for any variant of the interaction of the x-ray photon with the atom. For coherent scattering, the approximate calculations of the intensity of scattered radiation may be carried out using equations (5.26) and (5.32), and equation (5.36) may be used for noncoherent scattering. An increase of the photon energy is accompanied by the increase of the probability of noncoherent (Compton) scattering which becomes dominant at the photon energies of E>200 keV (ie, in the region of γ-radiation) for any atomic numbers of the scattering material. An increase of the amount by which the photon energy exceeds the binding energy of the electron in the atom increases the efficiency of the description of the scattering process by the previously examined theoretical expressions for noncoherent scattering on free electrons. Figure 5.10 shows the dependences of the atomic coefficient σ_a of coherent and noncoherent scattering on the energy of the photon of scattering radiation for iron atoms. The calculations were carried out using the tabulated data from [218].

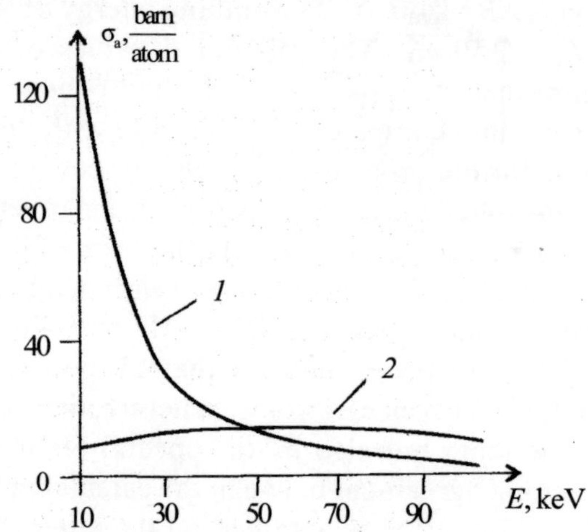

Fig. 5.10. Dependence of the atomic coefficients of coherent (curve 1) and noncoherent (curve 2) scattering on the photon energy of scattering radiation for iron atoms.

Figure 5.10 shows that for the iron atoms at the photon energy exceeding approximately 50 keV the non-coherent scattering becomes dominant in relation to coherent scattering.

The relationship of the components of the processes greatly depends on the atomic number of the scattering element.

The overall scattering coefficient decreases with increasing photon energy, although the rate of decrease is smaller in comparison with the coefficient of photoelectric absorption. This is indicated by the data in Fig. 5.11 on the interaction of high energy photons with lead atoms.

Figure 5.11 shows that in the examined case, the dominant role in the attenuation of radiation with the energy lower than 300 keV is played by the process of photoelectric absorption. At 500 keV, the probabilities of photoelectric absorption and Compton scattering in a lead specimen are approximately identical. At energies from 0.7 to 4 MeV the main factor in the attenuation of radiation is the scattering process. At energies higher than 4 MeV the controlling process in attenuation of the photon beam is the process of formation of the electron–positron pair in the electric field of the nucleus [219, 220].

5.3. Intensity of x-ray radiation scattered by a thick specimen

To examine the processes of scattering of x-ray radiation by a thick specimen it is necessary to pass from the atomic coefficients, expressed by the equation (5.27) for coherent and (5.35) for non-coherent radiation, to linear $d\sigma_x/d\vartheta$ and mass $d\sigma_m/d\vartheta$ scattering coefficients. This transition is

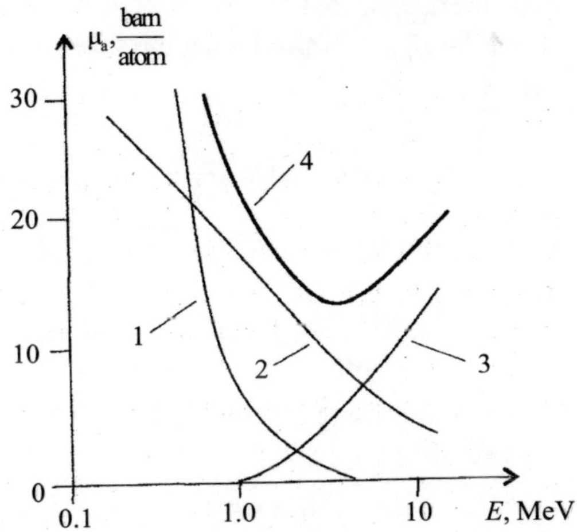

Fig. 5.11. Dependence of the components of the atomic coefficient of attenuation of x-ray and gamma radiation μ_a on the photon energy for lead. 1) photoelectric (true) absorption τ, 2) scattering σ, 3) formation of electron–positron pairs in the electrical field of the nucleus, 4) the overall attenuation coefficient.

identical with that in the case of the absorption coefficients:

$$\frac{d\sigma_x}{d\Omega} = \frac{\rho N}{A}\frac{d\sigma_a}{d\Omega} \tag{5.40}$$

and

$$\frac{d\sigma_m}{d\Omega} = \frac{N}{A}\frac{d\sigma_a}{d\Omega}, \tag{5.41}$$

where N is the Avogadro number, A is the atomic weight of the element, and ρ is the density of matter.

The mass scattering coefficient for the materials with a complicated composition (like the mass absorption coefficient) is determined by the sum of the mass scattering coefficients of the components of this material, multiplied by their content C_i:

$$\frac{d\sigma_m}{d\Omega} = \sum_i C_i \frac{d\sigma_{mi}}{d\Omega}. \tag{5.42}$$

These relationships are valid for both differential and total coefficients of coherent and non-coherent scattering.

To derive the expression for the intensity of x-ray radiation, scattered by a thick specimen, in [31], the authors investigated the flat surface of such a specimen, with a parallel beam of x-ray radiation with the wavelength λ_1 and intensity I_0 falling on the surface under the angle φ. At the depth x the radiation is scattered in the elementary layer dx under the angle ψ to the surface of the specimen (Fig. 5.12)

The intensity I_x at the depth x is linked with the intensity I_0 on the surface of the specimen by the equation:

$$I_x = I_0 \exp\left\{-\mu(\lambda_1)\frac{x}{\sin\varphi}\right\}.$$

Attenuation of intensity in the layer dx is determined as follows:

$$dI_x = -I_0 \frac{\mu(\lambda_1)}{\sin\varphi} \exp\left\{-\frac{\mu(\lambda_1)}{\sin\varphi}x\right\}dx.$$

Since the primary beam is incident on the surface under the angle φ, the energy, attenuated by the area s is:

$$dE_x = dI_x s \sin\varphi,$$

or

$$dE_x = -I_0 s\mu(\lambda_1) \exp\left\{-\frac{\mu(\lambda_1)}{\sin\varphi}x\right\}dx.$$

The intensity of radiation (energy per unit area), scattered in the range of the solid angle $d\Omega$, at the distance $R \gg x$ from the specimen is determined from the equation:

$$dI_p = \frac{dE_x}{R^2} = \frac{I_0}{R^2} s \frac{d\sigma_x}{d\Omega} \exp\left\{-\frac{\mu(\lambda_1)}{\sin\varphi}x\right\}dx.$$

The minus sign is omitted because the part of the energy of primary radiation, lost in scattering, is acquired by scattered radiation.

Taking into account the attenuation of scattered radiation at exit from the specimen, the equation for dI_p, is transformed to the form:

$$dI_p = \frac{I_0}{R^2} s \frac{d\sigma_x}{d\Omega} \exp\left\{-\left(\frac{\mu(\lambda_1)}{\sin\varphi} + \frac{\mu(\lambda)}{\sin\psi}\right)x\right\}dx,$$

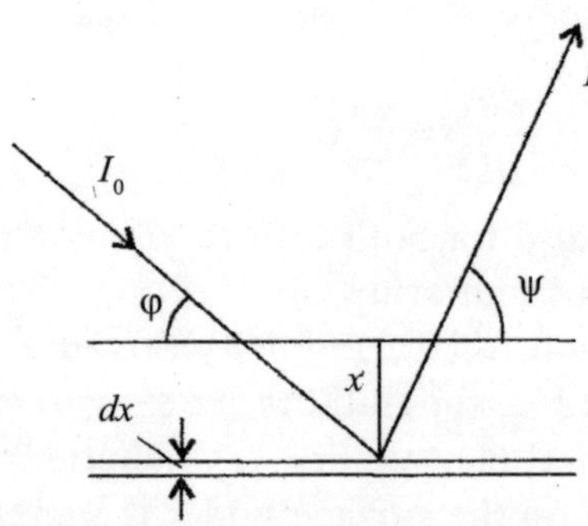

Fig. 5.12. Derivation of the equation for the intensity of scattered radiation.

where λ is the wavelength of scattered radiation, ψ is the angle of exit of scattered radiation from the specimen.

Integrating the resultant equation with respect to the depth x_i in the range from 0 to x gives:

$$I_p = \frac{I_0}{R^2} s \frac{d\sigma_x}{d\Omega} \frac{1 - \exp\left\{ -\left(\frac{\mu(\lambda_1)}{\sin\varphi} + \frac{\mu(\lambda)}{\sin\psi} \right) x \right\}}{\frac{\mu(\lambda_1)}{\sin\varphi} + \frac{\mu(\lambda)}{\sin\psi}}. \tag{5.43}$$

Transferring from the linear absorption and scattering coefficients to the mass coefficients:

$$I_p = \frac{I_0}{R^2} s \frac{d\sigma_m}{d\Omega} \frac{1 - \exp\left\{ -\left(\frac{\mu_m(\lambda_1)}{\sin\varphi} + \frac{\mu_m(\lambda)}{\sin\psi} \right) p_s \right\}}{\frac{\mu_m(\lambda_1)}{\sin\varphi} + \frac{\mu_m(\lambda)}{\sin\psi}}, \tag{5.44}$$

where $p_s = \rho x$ is the surface density of matter.

In the case of thick specimens ($p_s \to \infty$), equation (5.44) gives:

$$I_p = \frac{I_0}{R^2} \frac{d\sigma_m}{d\Omega} \frac{s}{\frac{\mu_m(\lambda_1)}{\sin\varphi} + \frac{\mu_m(\lambda)}{\sin\psi}}. \tag{5.45}$$

For high- and medium-wavelength values, when the difference of the wavelength of the coherently and non-coherently scattered radiation may be ignored, equation (5.45) gives:

$$I_p \approx \frac{I_0}{R^2} \frac{s}{\left(\frac{1}{\sin\varphi} + \frac{1}{\sin\psi} \right)} \frac{1}{\mu_m(\lambda_1)} \frac{d\sigma_m}{d\Omega}. \tag{5.46}$$

In the investigated case, the mass scattering coefficient $d\sigma_m/d\Omega$ is the sum of the coherent and non-coherent differential coefficients for the radiation with wavelength λ_1.

Equation (5.46) shows that the intensity of scattered radiation is proportional to the differential mass scattering coefficient and inversely proportional to the mass coefficient of attenuation of scattered radiation. The absorption coefficient increases in proportion to Z^4 and the scattering coefficient depends to a lesser extent on the atomic number Z. Therefore, the intensity of radiation, scattered by the thick specimen, decreases with the increase of the atomic number Z of the scattering specimen. To confirm this conclusion, Fig. 5.13 shows the spectral distribution of the radiation of an x-ray tube with a molybdenum anode in scattering by aluminium and copper specimens [31].

Figure 5.13 shows that the intensity of the radiation, scattered by the aluminium specimen ($Z = 13$), is considerably higher than the intensity for the copper specimen ($Z = 29$) of the same size.

Figure 5.13 shows the distinctive doublet of coherently scattered lines Mo K$\alpha_{1,2}$. The width of the doublet is determined by the natural width of the line and by the resolution of the spectrometer. The width of the non-coherently scattered lines of the same doublet is considerably greater. The increase of the width of the noncoherent scattering lines was discussed previously and is determined by the bonds of the electrons in the atom and by the movements of the electrons in the field of the nucleus.

In addition, for the shortwave x-ray radiation, the broadening of the line is determined by the process of multiple noncoherent scattering. According to [221], the contribution of this process to the total scattering intensity in monochromatic radiation of the source Cd-109 (photon energy 22.8 and 88 keV) in the case of interaction with an aluminium specimen was 14% at 22.8 keV and 60% at 88 keV. The energy distribution of the intensity of noncoherently scattered radiation of the source at the geometry $\varphi = 45°$ and $\psi = 90°$ is shown in Fig. 5.14.

Figure 5.14 shows that multiple non-coherently scattered radiation is superimposed on the peak of the single non-coherently scattered radiation. This is clearly evident for the non-coherent line with the energy of 88 keV.

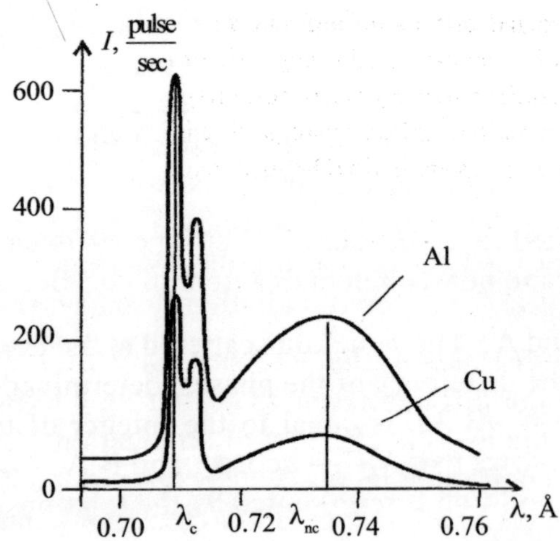

Fig. 5.13. Spectral distribution of the radiation of an x-ray tube with a molybdenum anode. Scattering takes place on copper and aluminium specimens.

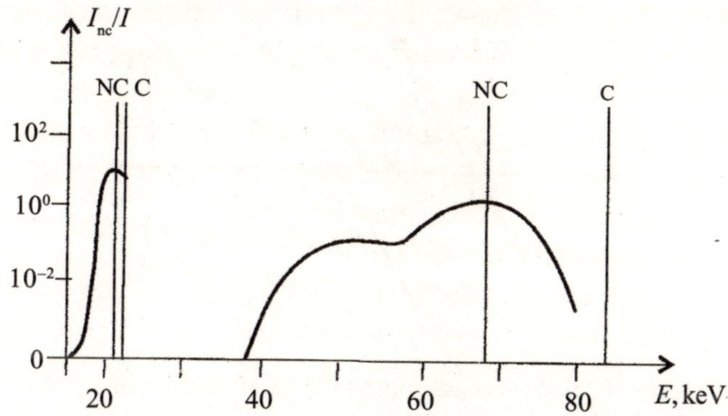

Fig. 5.14. Energy distribution of the radiation of Cd–109 source non-coherently scattered on the aluminium specimen ($\varphi = 45°$, $\psi = 90°$). The ratio I_{nc}/I for the maximum of non-coherently scattering radiation with the energy of 88 keV is equal to unity (I is the intensity of primary radiation).

5.4. Scattering of x-ray radiation by ordered structures

The previously described equation (5.46) for the intensity of the scattered x-radiation of the thick specimen does not take into account the distribution of the atoms in the scattering object. This is valid for the noncoherently scattered x-radiation, with no interference. The coherently scattered radiation, strictly speaking, does not interfere only in the case of the monatomic gases where the distances between the atoms are considerably greater than the size of the atom, and the atoms themselves move randomly and, consequently, the radiation coherently scattered on the individual atoms is not linked in phase. In the examination of the scattering of x-radiation on an ideal crystal [222] it is taken into account that this crystal has an ordered three-dimensional lattice. In the lattice, one of the atomic rows A_1, A_2, A_3,..., is separated. In this row, the atoms are situated at identical distances *a* from each other (Fig. 5.15). The angles between the selected atomic row and the primary parallel beam are denoted by α_0, and α_p is the angle between the row and the parallel beam of scattering radiation.

It is assumed that the coherent radiation forms on the orbital electrons of the atoms A_1 and A_2. The radiation scattered at the points A_1 and A_2 will interfere only if the difference of the phases, determined by the difference of the sections $A_1C_1-B_2A_2$, is equal to the integer of the wavelength λ. Figure 5.15 also shows that $A_1C_1 = a\cos\alpha_p$, and $B_2A_2 = a\cos\alpha_0$. Therefore, the interference condition is represented by the relationship:

$$a(\cos\alpha_p - \cos\alpha_0) = p\lambda, \qquad (5.47)$$

where p is the integer.

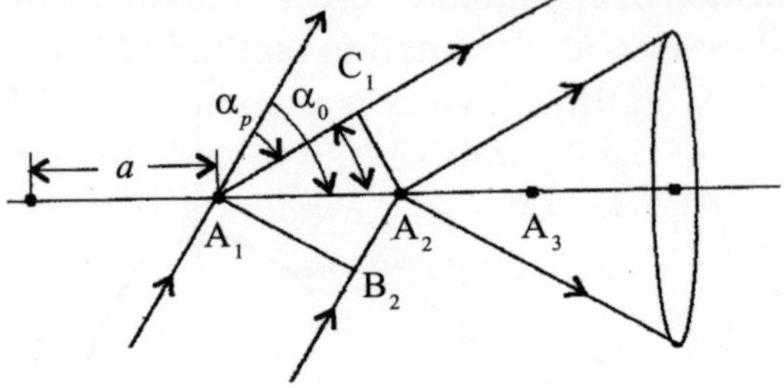

Fig. 5.15. Diagram of coherent scattering of the parallel beam of x-radiation by an ordered row of atoms A_1, A_2, A_3,...

The resultant condition is valid for all directing cones with the tips at the points A_1, A_2,... and with the angle at the tip of $2\alpha_p$ (the figure shows the one of these cones).

Writing a similar relationship for the three main nodal rows, parallel to the spatial coordinates axes X, Y, Z, leads to the Laue condition, derived in 1912:

$$\begin{aligned} a(\cos\alpha_p - \cos\alpha_0) &= p\lambda, \\ b(\cos\beta_q - \cos\beta_0) &= q\lambda, \\ c(\cos\gamma_r - \cos\gamma_0) &= r\lambda, \end{aligned} \tag{5.48}$$

where a, b, and c are the periods of identity along the axes X, Y, Z, α_0, β_0 and γ_0 are the angles between these axes and the direction of the primary beam, α_p, β_q and γ_r are the angles between these axes and the direction of the interference beam, p, q, and r are integers.

The interference in the conditions of the three-dimensional lattice requires that all the three relationships (5.48) are fulfilled. However, these relationships are independent of each other and the directions which satisfy them simultaneously do not exist. In other words, for the given values of p, q and r, all the three cones, similar to the one shown in Fig. 5.15, do not intersect in the same direction. This intersection, i.e., the interference effect for the spatial lattice, may be realised either by selecting the wavelength λ of the scattered radiation or by changing the orientation of the lattice (by varying the parameters α_0, β_0 and γ_0).

If the primary radiation is polychromatic, its wavelengths include such wavelengths which produce cones intersecting in the same direction. In particular, this method of production of the interference pattern was realised by Friedrich and Knipping [223] following Laue's proposal. It is therefore referred to as the Laue method.

If a single crystal is rotated in the flux of monochromatic radiation, then under certain dispostion of crystal interference maxima will also be detected. This method is referred to as the rotation method [224].

Finally, if a polycrystalline powder is used instead of the single crystal, then the large group of the crystals should include the crystals oriented in relation to the primary beam of monochromatic radiation in such a manner that all the interference conditions are detected at the same time. This method is referred as the powder method or the Debye–Scherrer method [225].

The process of coherent scattering on ordered structures with subsequent interference of the x-ray radiation may be formally regarded as the reflection of radiation from the atomic planes.

It is assumed that a flat wave of x-radiation falls on two adjacent atomic planes of the crystal under the angle φ. The difference in the path of the beams (Fig. 5.16), scattered on the orbital electrons of these atoms, is determined by the value BA_2C and may be expressed by the planar spacing d as $BA_2C = 2d \sin\varphi$. If this path difference is equal to the integer wavelength, the scattered radiation will interfere. Taking this into account we obtain a law established by W.L and W.H. Bragg [226] in England and independently by G. Wulf [227] in Russia, presented in the form:

$$2d \sin\varphi = n\lambda. \tag{5.49}$$

The Wulf–Bragg law indicates that in the case of polychromatic primary radiation, incident on the crystal under angle φ, only the waves whose wavelength satisfies the condition (5.49) are reflected from the single crystals. If the incident radiation is monochromatic, the Wulf–Bragg law gives the required orientation of the single crystal to ensure reflection from a specific set of the atomic planes.

Real single crystal are mosaic to a certain degree, i.e., the position of the individual blocks and atoms in this crystals deviates from the mean position. This deviation $\delta\varphi$ is usually 10^{-2}–10^{-4} rad. The mosaic nature

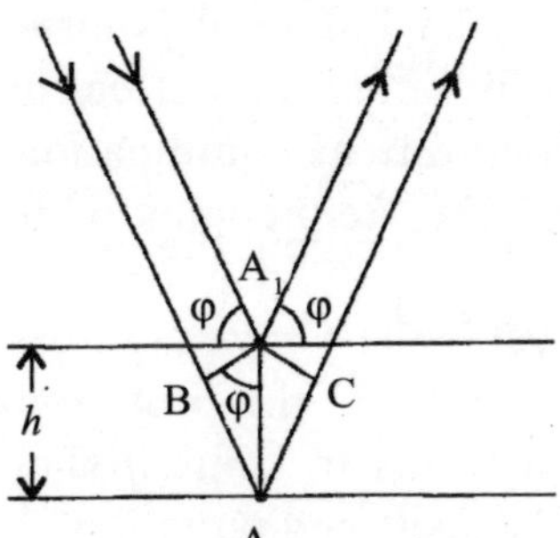

Fig. 5.16. Interference of x-ray radiation as a reflection from atomic planes in an ideal crystal.

results in the expansion of diffraction maxima, and the reflection coefficient equals 0.1–0.4 of the case of ideal crystals. However, the integral intensity (the area below the diffraction curve) of the mosaic crystals is higher because it is determined by the product of the extent of the mosaic shape by the reflection coefficient.

The great majority of the powder materials, metals, alloys and other condensed media are fine-crystalline formations. The individual blocks of these materials (crystals) are randomly oriented in space. Therefore, for a collimated beam of monochromatic x-ray radiation, the large group of the crystals always contains crystals oriented under the Wulf–Bragg angle in relation to the beam. This is accompanied by the unavoidable formation of diffraction maxima of the first, second, and higher orders of reflection. The maxima are detected on the background of diffusely scattered radiation. The ratio of the intensities of the maxima and the background depend on the size of the crystals. As the size of the crystal decreases, the intensity of the background increases.

Study of the coherent scattering of x-ray radiation on the ordered atomic structures is one of the main means of crystallographic investigations and forms the basis of x-ray diffraction analysis used widely in the examination of the properties and characteristics of natural and industrial materials.

5.5. Resonance combination scattering of x-ray radiation

Resonance combination scattering is determined by completely different processes in comparison with the previously investigated coherent and non-coherent scattering.

For the electromagnetic radiation of the optical range, the effect of combination scattering, associated with the variation of the frequency of this radiation to the frequency of natural oscillations of the scattering particles, was discovered in 1928 by Landsberg and Mandelstam [228, 229] for crystals and independently by Indian physicists Raman and Krishnan [230] for liquids. In foreign literature, this effect is referred to as Raman scattering, and in Russian literature as combination scattering, because it represents a combination of the frequencies of the forced and natural oscillations.

The Raman scattering effect in the x-ray range was discovered by Sparks in 1974 [231, 232]. In this effect, at the proton energy E_1, close to the binding energy E_q of the electrons in some q-shell, insufficient for the ionisation of the shell ($E_1<E_q$), scattered radiation photons appear with the energy smaller than E_1 by the binding energy of the electronic level, situated above the resonance level.

The theory of Raman scattering was described in [233–235]. In the

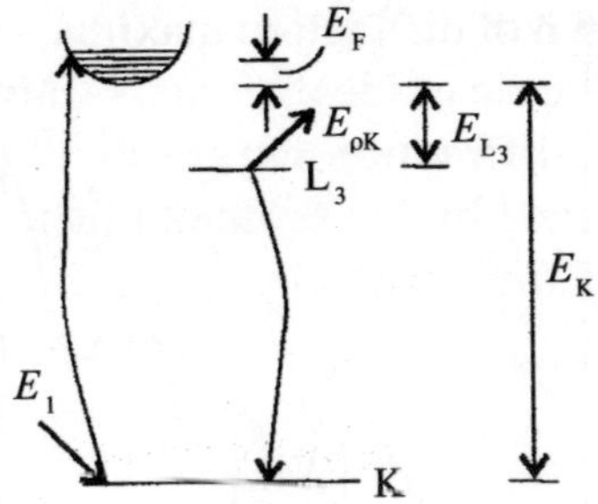

Fig. 5.17. Diagram of formation of the line of resonance Raman scattering.

first approximation, the theory may be described as follows:

An electron, for example, of the K-shell (Fig. 5.17, taken from [31]) is transferred under the effect of a photon with the energy $E<E_K$ to the conduction band outside the limits of the atom because of the combination of the amplitude of the forced and natural oscillations of the electron. The resultant K-vacancy is filled with an electron of a higher shell (for example, L_3-shell). This is accompanied by the emission of a quantum of combination (Raman) scattering with the energy E_c equal to:

$$E_c = E_1 - (E_{L3} + E_F), \tag{5.50}$$

where E_F is the energy of the Fermi level to which the conduction band is filled. The electron may also be ejected to free discrete levels of the atomic-molecular system [235–237].

Equation (5.50) shows that the energy of Raman scattering of the photon depends on the energy E_1 of scattered radiation. If this radiation is monochromatic, lines of resonance Raman scattering appear. These lines are non-symmetric and contain a tail on the side of low energies. Every characteristic line of incident radiation is linked with the corresponding line of Raman scattering. If the primary photons belong to $K\alpha_1$ line of some element ($E_1 = E_{K\alpha}$), then for the investigated case of the $K \rightarrow L_3$ transition in the scattering element the energy shift of the resonance line in relation to $E_{K\alpha}$ (or of the coherently scattered peak of this line) is, according to (5.50), approximately equal to E_{L3}, since the energy E_F of the Fermi level is considerably smaller than E_{L3}.

Resonance Raman scattering radiation is isotropic, independent of the scattering angle.

Figure 5.18 show the experimentally determined [238] energy distribution of Mo $K\alpha$-radiation, scattered by a zirconium specimen. The energy of this radiation ($E_{Mo\,K\alpha1}$=17.479 keV) is insufficient for the photoionisation of the K-shell of the zirconium atoms ($E_{K\,Zr}$=17.999 keV). At the same time, the compared energies are similar and this results in Raman scattering.

Figure 5.18 shows that the line of resonance Raman scattering (rs) on

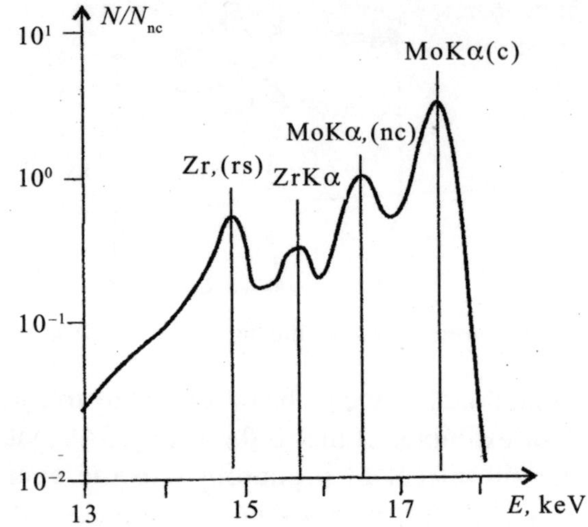

Fig. 5.18. The energy spectrum of scattering of Mo Kα radiation by zirconium.

the zirconium atoms is detected on the background of non-coherent scattering (nc) of molybdenum radiation. The appearance of the fluorescent Zr Kα-line is determined by the presence, in the composition of primary radiation, of a short-wave bremsstrahlung component, formed as a result of its diffusion scattering on the crystal–monochromator used for the separation of the Mo Kα-line from the polychromatic radiation of the x-ray tube.

Raman scattering forms in the area of the absorption edges, i.e. coincides with the region of the anomalous coherent scattering (see section 5.3.1) of x-ray radiation. The integral atomic coefficient of resonance Raman scattering σ_a^{rs} is proportional to the correction ΔF for anomalous coherent scattering [232, 238]:

$$\sigma_a^{rs} = 4\pi r_0^2 \Delta F, \tag{5.51}$$

where r_0 is the classic electron radius (see section 5.2).

Equation (5.51) shows that the decrease of the intensity of coherent scattering in the vicinity of the absorption edges, proportional to the value $(F-\Delta F)^2$, is accompanied by an increase of the intensity of resonance Raman scattering.

The Raman scattering lines are sensitive to changes in the chemical bonds of the atoms. Usually, the energy shift of these lines is investigated in relation to more stable lines of x-ray fluorescence [235, 239]. The strongest effect of the chemical bonds is detected in transitions of the electrons during absorption on non-filled outer shells of the atoms [236, 237] because these outer shells take part in atomic interactions. Not only the shift of the appropriate lines but also the very fact of the appearance of these lines is important.

Chapter 6

Refraction and reflection of x-ray radiation

According to the electromagnetic theory of light, the effect of the electromagnetic wave on the irradiated specimen results in the formation of forced oscillations of the electrons leading to the formation of an induced wave. The superposition of the induced wave onto the primary wave results in the formation of the resultant wave which propagates in a changed direction, i.e., the transition from one medium to another is accompanied by the refraction of electromagnetic radiation. Refraction takes place at any wavelength of the radiation incident on the specimen, and all the relationships of this interaction for the visible range of the spectrum [240, 241] are also valid for x-ray radiation.

In particular, in the x-ray range, the refraction of the beam at the interface of two media is described by Snell's law:

$$V_0 \cos\varphi = V \cos\varphi_0, \tag{6.1}$$

where φ_0 and φ are the sliding angles for the incident and refracted radiation, V_0 and V are the phase velocities of the propagation of radiation in the contacting media.

Equation (6.1) shows that:

$$n_0 \cos\varphi_0 = n \cos\varphi, \tag{6.2}$$

where $n_0 = c/V_0$ and $n = c/V$ are the refractive indexes for the contacting media, c is the velocity of light.

For the relative refractive index:

$$\frac{n}{n_0}=\frac{V_0}{V},$$

or, taking into account that for the medium $\lambda = V/\nu$, and the frequency of ν radiation does not change in transition from one medium to another, gives:

$$\frac{n}{n_0}=\frac{\lambda_0}{\lambda}. \tag{6.3}$$

If the radiation is incident on the surface of the medium from vacuum, then $n_0 = 1$, and for the refractive index n (referred to as the absolute refractive index in this case):

$$n=\frac{\lambda_0}{\lambda}. \tag{6.4}$$

Or, using, (6.2), for the absolute refractive index of the medium:

$$n=\frac{\cos\varphi_0}{\cos\varphi}. \tag{6.5}$$

In the optical wavelength range, any medium is optically denser than vacuum, i.e., $n > 1$. Consequently, $\varphi_0 < \varphi$ and in incidence of radiation from vacuum on the specimen, the direction of the reflected beam is closer to the normal than the direction of the incident beam.

The optical properties of x-ray radiation have a number of special features. To determine these features, attention will be given initially to the dependence of the refractive index on the wavelength of incident radiation. This dependence is referred to as dispersion.

6.1. Theoretical fundamentals of studying dispersion

The dispersion in the range of x-ray radiation is examined in detail in a monograph by Blokhin [7].

The partial refractive index n_q for the electrons of the q-level according to the Lorenz theory is described by the expression:

$$n_q = 1-\delta_q - i\beta_q, \tag{6.6}$$

where β_q is determined by the absorption properties of the specimen and is associated with the partial absorption coefficient τ_q by the equation:

$$\beta_q = \frac{\lambda}{4\pi}\tau_q, \tag{6.7}$$

and the unit decrement of the partial refractive index δ_q is determined from the equation:

$$\delta_q = \frac{2\pi e^2 l_q}{m} \frac{\omega^2 - \omega_q^2}{(\omega^2 - \omega_q^2)^2 + 4b_\omega^2 \omega^2}. \tag{6.8}$$

In (6.8), l_q is the number of the q-electrons in the unit volume of the medium, ω is the circular frequency of forced oscillations, ω_q is the circular frequency of intrinsic oscillations of the electrons of the q-shell in the atom:

$$\omega_q = \frac{2\pi c}{\lambda_q} = 2\pi\nu_q,$$

b_ω is a quantity characterising the process of attenuation of forced oscillations:

$$b_\omega = \frac{e^2}{3mc^2}\omega^2.$$

No attention will be given to the range of frequencies ω close to the frequency of intrinsic oscillations ω_q. Let the value $(\omega^2 - \omega_q^2)^2$ be so high that $(\omega^2 - \omega_q^2)^2 \gg 4 b_\omega^2 \omega^2$. In this case, from equation (6.8):

$$\delta_q = \frac{2\pi e^2}{m} \frac{l_q}{\omega^2 - \omega_q^2} = \frac{e^2}{2\pi m} \frac{l_q}{\nu^2 - \nu_q^2}. \tag{6.9}$$

Thus, summing up with respect to all shells of the atom:

$$\delta = \frac{e^2}{2\pi m} \sum_q \frac{l_q}{\nu^2 - \nu_q^2}. \tag{6.10}$$

The theory, proposed by Lorenz, ignores the absorption, assuming that the value of β_q in equation (6.6) is equal to zero. This assumption is substantiated by the data published in Table 6.1 which compares the values of δ and β for the shortwave characteristic Mo Kα-radiation. (For the compounds, presented in the table, $\delta = \sum_i c_i \delta_i$ and $\beta = \sum_i c_i \beta_i$, where c_i is the content of the i–th element in the compound).

Table 6.1 shows that even for gold (Z = 79), the value of δ is almost eight times higher than the value of β.

In the optical range of the spectrum, where $\nu \ll \nu_q$, according to the equation (6.10) $\delta < 0$. For the x-ray range where $\nu \gg \nu_q$, the decrement of the refractive index is positive ($\delta > 0$).

Consequently, taking into account the equation

$$n = 1 - \delta, \tag{6.11}$$

It may be seen that the refractive index n for the x-ray range of the spectrum is smaller than unity.

In other words, for x-ray radiation any medium is optically less dense

Table 6 .1. Values of δ and β of various refracting media for x-ray Mo Kα-radiation (λ = 0.71 Å)

Medium	$\delta \cdot 10^8$	$\beta \cdot 10^8$
Plexiglass	90	0.055
Quartz glass	150	0.46
Silicon	160	0.84
Copper	560	24.1
Tantalum	910	87.5
Gold	1050	129.5

than vacuum. Therefore, at $n=\frac{\cos\varphi_0}{\cos\varphi}<1$ and $\varphi<\varphi_0$ (the refracted x-ray beam forms a smaller angle with the surface than the incident beam).

Assuming for the x-ray range of the spectrum that $\nu \gg \nu_q$, from equation (6.10):

$$\delta = \frac{e^2 L}{2\pi mc^2}\lambda^2, \tag{6.12}$$

where $L=\sum_q l_q$ is the number of electrons in the unit volume, λ is the wavelength of x-ray radiation.

Since the number of electrons in the unit volume is determined by the expression:

$$L = \frac{\rho N Z}{A},$$

then

$$\delta = \frac{e^2 N}{2\pi mc^2}\left(\frac{\rho Z}{A}\right)\lambda^2 = 2.70\cdot 10^{10}\frac{\rho Z}{A}\lambda^2, \tag{6.13}$$

where λ is expressed in centimetres.

From equation (6.13):

1. The value of δ is very low. For example, in interaction of radiation of Mo Kα (λ = 0.71 Å) with aluminium, the value of δ according to calculations is equal to $\delta = 1.74\cdot 10^{-6}$. This value is close to the experimentally determined value of δ.

2. For a given substance (i.e., when the value of L is constant), the ratio $\frac{\delta}{\lambda^2}$ is constant: $\frac{\delta}{\lambda^2}=\frac{e^2 L}{2\pi mc^2}$=const. The constancy of $\frac{\delta}{\lambda^2}$ is also efficiently

confirmed by the experiments for $\lambda \ll \lambda_q$.

The previously mentioned proximity of the refractive index to unity creates difficulties in solving a number of problems in practice. To increase the value of the index, it would be necessary to use larger wavelengths. However, the increase of λ moves the wavelength to the range of the wavelengths λ_q of the absorption edges where the dispersion is complicated by the effect of the frequency of intrinsic oscillations of the electrons of the atom and by the large increase of the true absorption coefficient τ increasing the value of β.

The suitability of the Lorenz theory for the so-called range of anomalous dispersion, in which λ is close to λ_q, will be evaluated. Passing in equation (6.10) from frequencies to wavelengths, for δ/λ^2:

$$\frac{\delta}{\lambda^2} = \frac{e^2}{2\pi mc^2} \sum \frac{l_q \lambda_q^2}{\lambda_q^2 - \lambda^2}. \qquad (6.14).$$

Expressing l_q (the number of q-electrons in the unit volume) by means of z_q – the number of q-electrons in the atom, where λ_q is close to λ:

$$l_q = \frac{\rho N}{A} z_q.$$

Here ρ is the density of matter, A is the atomic weight, N is the Avogadro number.

Substituting this equation into (6.14) and separating the K-shell from the sum gives:

$$\frac{\delta}{\lambda^2} = \frac{e^2}{2\pi mc^2} \frac{\rho N}{A} \left(\frac{z_K \lambda_K^2}{\lambda_K^2 - \lambda^2} + \sum_{q \neq K} \frac{z_q \lambda_q^2}{\lambda_q^2 - \lambda^2} \right). \qquad (6.15)$$

The range of the wavelength in the vicinity of the wavelength of the K-adsorption edge will be examined. In this case, $\lambda \ll \lambda_q$ for all the remaining shells, and it may be assumed that:

$$\sum_{q \neq K} \frac{z_q \lambda_q^2}{\lambda_q^2 - \lambda^2} \approx \sum_{q \neq K} z_q = Z_A - z_K.$$

Taking into account the resultant equation, equation (6.15) is presented in the following form:

$$\frac{\delta}{\lambda^2} = \frac{e^2}{2\pi mc^2} \frac{\rho N}{A} \left(Z_A - \frac{z_K}{1 - \frac{\lambda_K}{\lambda}} \right). \qquad (6.16)$$

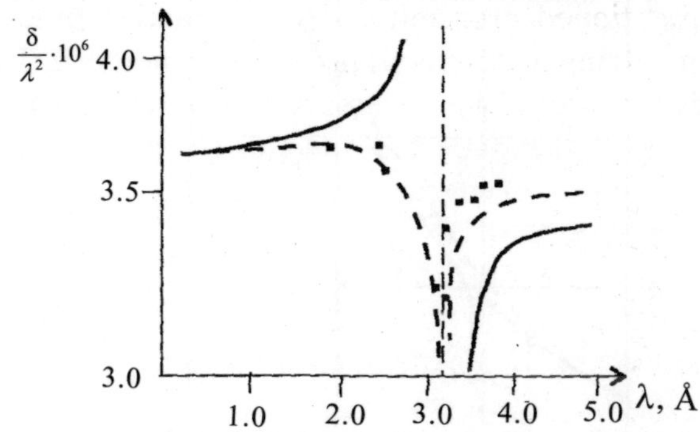

Fig. 6.1. Dependence of δ/λ^2 on the wavelength λ for calcite in the range of the K-absorption edge of calcium. Solid line – calculated according to Lorenz. The broken line – calculated according to Kallman and Mark. The points indicate the experimental results.

Equation (6.16) shows that at $\lambda = \lambda_K$, an increase of λ results in the change of the function δ/λ^2 from $+\infty$ to $-\infty$.

The values calculated from equation (6.16) and the experimental values δ/λ^2 for calcite ($CaCO_3$) are compared in Fig. 6.1.

Figure 6.1 shows that the Lorenz theory, which is fully applicable in the range of the short wavelengths, gives too low results in the range $\lambda > \lambda_K$, and is not accurate in the vicinity of the absorption edge. The reason for this discrepancy between the calculated and experimental results is that the absorption of incident radiation by the medium is ignored. This absorption of radiation in the range λ_q is, as previously mentioned, significant.

The effect of the absorption is taken into account in studies by Kallman and Mark [242, 243]. It has been shown that the value δ_q/λ^2 for the q-shell is determined by the expression:

$$\frac{\delta_q}{\lambda^2} \approx \frac{e^2 l_q}{2\pi mc^2}\left(1+\frac{\lambda^2}{\lambda_q^2}\ln\left|1-\frac{\lambda_q^2}{\lambda^2}\right|\right).$$

Summing up with respect to all shells of the atom, they obtained:

$$\frac{\delta}{\lambda^2} \approx \frac{e^2}{2\pi mc^2}\left(\sum_q l_q + \lambda^2 \sum_q\left(\frac{l_q}{\lambda_q^2}\ln\left|1-\frac{\lambda_q^2}{\lambda^2}\right|\right)\right). \tag{6.17}$$

Comparing equation (6.17) with the equation determined from the Lorenz theory (equation (6.14)), it may be seen that instead of the sum which contains the difference of the squares $\lambda_q^2 - \lambda^2$, in the denominator, a new more complicated equation is obtained. At $\lambda = \lambda_q$, the ratio δ/λ^2 becomes equal to $-\infty$. The part of the curve for the range $\lambda < \lambda_K$ in the previously

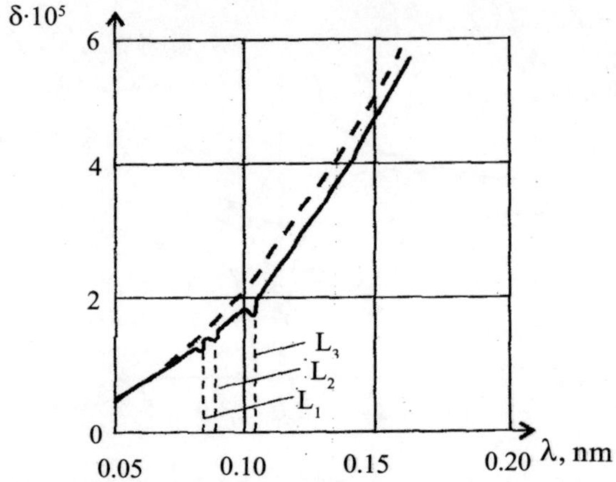

Fig. 6.2. Dependence of the decrement δ on the wavelength of incident radiation for gold ($Z = 79$) (solid curve). The dotted curve indicates the results of approximate calculations using equation (6.13).

investigated case (see Fig. 6.1) of reflecting of radiation on calcite is in satisfactory agreement with the experimental data. The remaining small discrepancy between the theory and experiment for the range $\lambda > \lambda_K$ is explained by the fact that screening of the electrons of the K-shell by other electrons of the atom has not been taken into account.

The fact that this screening is taken into account in the quantum theory of dispersion, proposed by Hohl [244], resulted in good agreement also in the wavelength range behind the absorption edge.

Similar areas of anomalous dispersion of x-ray radiation are also detected in the vicinity of the wavelength of the L-, M- and other absorption edges.

The results of accurate and approximate calculations of the dependence of decrement δ on the wavelength λ of incident radiation in the anomalous range of the L-absorption edge for the surface of gold are presented in Fig. 6.2.

6.2. Refraction of x-rays

The refractive index in transition of radiation from vacuum into some medium is determined by the equation (6.4)

$$n = \frac{\cos\varphi_0}{\cos\varphi},$$

which in the range of x-ray radiation, as shown in the previous section, is lower than unity ($n<1$). Consequently, the angle of incidence φ_0 of radiation is greater than the angle φ under which radiation propagates in the medium

(φ_0>φ). Therefore, there is a minimum angle $(\varphi_0)_{min}$ at which radiation does not penetrate from vacuum into matter. As indicated by Fig. 6.3, in this case $\Delta\varphi = (\varphi_0)_{min}$.

The dependence of $\Delta\varphi$ on the angle φ at higher values of the angle will be determined. Using the equations (6.11) and (6.4):

$$n = \frac{\cos\varphi_0}{\cos\varphi} = 1 - \delta.$$

Since δ is a small value, then

$$\frac{\cos\varphi_0}{\cos\varphi} = \frac{1-\delta^2}{1+\delta} \approx \frac{1}{1+\delta}. \tag{6.18}$$

From this $\cos\varphi = \cos\varphi_0 + \delta\cos\varphi_0$, or

$$\cos\varphi_0 - \cos\varphi = -\delta\cos\varphi_0. \tag{6.19}$$

If it is assumed that the angles φ_0 and φ differ only slightly from each other, then:

$$\Delta\cos\varphi = -\sin\varphi_0\Delta\varphi. \tag{6.20}$$

From the last two equations:

$$\Delta\varphi = \delta \operatorname{ctg} \varphi_0. \tag{6.21}$$

Equation (6.21) shows that:

1) at $\varphi_0 = \pi/2$ $\Delta\varphi = 0$, i.e., the radiation does not change its direction;

2) with a decrease of φ_0 from $\pi/2$ to $(\varphi_0)_{min}$, the value of $\Delta\varphi$ increases, reaching a maximum $(\Delta\varphi)_{max}$ at $\varphi_0 = (\varphi)_{min}$.

At low values of φ_0 close to $(\varphi)_{min}$, the values $\Delta\varphi$ and φ_0 are comparable. In these conditions, the differential transformation (6.20) cannot be used.

The deviation $\Delta\varphi$ for small values of φ_0 will be determined. At small values of φ_0, $\cos\varphi_0 = 1$ and in (6.19) $\delta\cos\varphi_0 \approx \delta$. Consequently

$$\cos\varphi_0 - \cos\varphi = -\delta. \tag{6.22}$$

The difference of the cosines is described by:

$$\cos\varphi_0 - \cos\varphi = -2\sin\frac{\varphi_0 - \varphi}{2}\sin\frac{\varphi_0 + \varphi}{2}.$$

Taking into account the small values of the angles, it may be assumed that the sinuses of these angles are equal to the angles. Consequently, $\cos\varphi_0 - \cos\varphi = -2\,(\Delta\varphi/2)\,(\varphi_0 + \varphi_0 - \Delta\varphi)/2$, or

$$\cos\varphi_0 - \cos\varphi = -\Delta\varphi\left(\varphi_0 - \tfrac{\Delta\varphi}{2}\right). \tag{6.23}$$

Substituting (6.23) into (6.22) gives: $\varphi_0\Delta\varphi - \Delta\varphi^2/2 - \delta = 0$ or $\Delta\varphi^2 - 2\varphi_0\Delta\varphi + 2\delta = 0$. Solving the resultant quadratic equation gives:

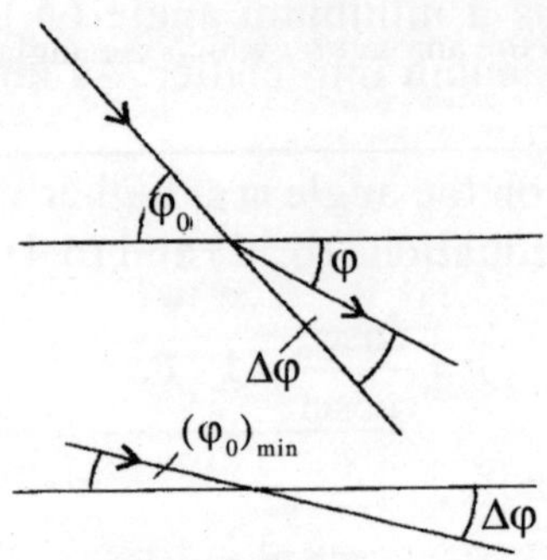

Fig. 6.3. The conventional and limiting variant of refraction of x-ray radiation.

$$\Delta\varphi = \varphi_0 - \sqrt{\varphi_0^2 - 2\delta}. \tag{6.24}$$

In the limiting case $(\Delta\varphi)_{max} = (\varphi_0)_{min} - \sqrt{(\varphi_0)_{min}^2 - 2\delta}$. Figure 6.3 shows that $(\Delta\varphi)_{max} = (\varphi_0)_{min}$. Separately, from this equation:

$$(\varphi_0)_{min}^2 = 2\delta,$$

or

$$(\varphi_0)_{min} = \sqrt{2\delta}. \tag{6.25}$$

Using (6.24) and (6.25) gives:

$$\Delta\varphi = \varphi_0 - \sqrt{\varphi_0^2 - (\varphi_0)_{min}^2}\,. \tag{6.26}$$

Equation (6.26) is suitable for examining the dependence of the deviation $\Delta\varphi$ on the value of the angle φ_0 and for comparing it with the identical dependence following from (6.21). It is assumed that the surface of aluminium (Z = 13, A = 27, ρ = 2.7 g/cm^3) receives radiation with a wavelength of 0.745 Å (Nb Kα-line). In this case, the region of dispersion is far from anomalous, and the value of δ may be calculated using the equation (6.13):

$$\delta = 2.70 \cdot 10^{10} \frac{\rho Z}{A} \lambda^2 \approx 2 \cdot 10^{-6}.$$

Therefore, in accordance with (6.25), $(\Delta\varphi)_{min}$:

$$(\varphi_0)_{min} = \sqrt{2\delta} = 2 \cdot 10^{-3}\ \text{rad}.$$

The values of the angle $\Delta\varphi$ for small incidence angles, calculated using equations (6.21) and (6.26), are presented in Table 6.2.

Table 6.2. Dependence of $\Delta\varphi$ on the amount by which the angle of incidence φ_0 exceeds the minimum angle $(\varphi_0)_{min}$

	$\varphi_0/(\varphi_0)_{min}$	5	2	1
$\Delta\varphi$[rad]	according to eq. (6.21)	$0.20\cdot10^{-3}$	$0.50\cdot10^{-3}$	$1.0\cdot10^{-3}$
	according to eq (6.26)	$0.20\cdot10^{-3}$	$0.54\cdot10^{-3}$	$2.0\cdot10^{-3}$

Table 6.2 shows that in the given conditions, the inaccuracy of the equation (6.21) is apparent only up to the angle φ_0 which is approximately twice the minimum refraction angle $(\varphi_0)_{min}$. In the case of large angles, the direction of the refracted beam is calculated using equation (6.21).

To complete this section, the approximate relationship between the angles φ and φ_0, used often at comparatively small values of these angles and taking into account the absorption of radiation by the refracting medium, will be derived. The absolute refractive index may be determined from the equation:

$$n = \frac{\cos\varphi_0}{\cos\varphi} = 1 - \delta - i\beta.$$

Squaring both parts of this equality gives:

$$\frac{1-\sin^2\varphi_0}{1-\sin^2\varphi} \approx 1 - 2(\delta + i\beta).$$

Assuming that the angles φ and φ_0 are small leads to:

$$\frac{1-\varphi_0^2}{1-\varphi^2} \approx 1 - 2(\delta + i\beta),$$

Taking into account the small value of the product $\varphi^2(\delta+i\beta)$, the above equation gives the approximate expression:

$$\varphi \approx \sqrt{\varphi_0^2 - 2\delta - 2i\beta}. \qquad (6.27)$$

6.3. Total external reflection of x-ray radiation

The minimum angle of sliding $(\varphi_0)_{min}$ at which x-ray radiation penetrates from vacuum into some medium, which was examined in the previous section, is the maximum angle of sliding $(\varphi_0)_{max}$ at which the radiation is reflected from the surface. This means that in this section it will be assumed that:

$$(\varphi_0)_{max} = \sqrt{2\delta}.$$

At $\varphi < (\varphi_0)_{max}$, there is total external reflection of x-ray radiation from the surface of the medium. This effect has been examined in detail in, for example, studies [7, 245, 246].

From the Frenel equations for the ratio of the electrical vectors of reflected and incident radiation it follows that the ratio of the intensities of these radiations is determined by the expression:

$$\frac{I}{I_0} = \frac{(\varphi_0 - a)^2 + b^2}{(\varphi_0 + a)^2 + b^2}, \tag{6.28}$$

where

$$a^2 = \frac{1}{2}\left(\sqrt{(\varphi_0^2 - 2\delta)^2 + 4\beta^2} + (\varphi_0^2 - 2\delta)\right),$$
$$b^2 = \frac{1}{2}\left(\sqrt{(\varphi_0^2 - 2\delta)^2 + 4\beta^2} - (\varphi_0^2 - 2\delta)\right). \tag{6.29}$$

If the medium is absolutely transparent for x-ray radiation, the quantity β, characterising absorption, is equal to zero. Consequently, from (6.29):

$$a = \sqrt{\varphi_0^2 - 2\delta} \quad \text{and} \quad b = 0.$$

In these conditions, the ratio of the intensities (6.28) is simplified and has the following form:

$$\frac{I}{I_0} = \left(\frac{\varphi_0 - \sqrt{\varphi_0^2 - 2\delta}}{\varphi_0 + \sqrt{\varphi_0^2 - 2\delta}}\right)^2 = \left(\frac{\varphi_0 - \sqrt{\varphi_0^2 - (\varphi_0)_{max}^2}}{\varphi_0 + \sqrt{\varphi_0^2 - (\varphi_0)_{max}^2}}\right)^2. \tag{6.30}$$

Equation (6.30) indicates that at $\varphi_0 = (\varphi_0)_{max}$ the intensity of reflected radiation is equal to the intensity of incident radiation. At $\varphi_0 > (\varphi_0)_{max}$ the ratio of the intensities rapidly decreases. The rate of this decrease, determined using equation (6.30), it is shown in Table 6.3.

Table 6.3 shows that at $\varphi_0 = 5(\varphi_0)_{max}$ the intensity of reflected radiation is equal to almost 0.

At $\varphi_0 < (\varphi_0)_{max}$ the equation (6.30) for the ratio of the intensities becomes complex. Therefore, in order to determine the ratio of the intensities in the given angle range φ_0, equation (6.30) is converted to the conventional complex form:

$$\sqrt{\frac{I}{I_0}} = \frac{\varphi_0 - i\sqrt{(\varphi_0)_{max}^2 - \varphi_0^2}}{\varphi_0 + i\sqrt{(\varphi_0)_{max}^2 - \varphi_0^2}},$$

or after multiplication of the numerator and the denominator by the difference of the terms of the denominator:

Table 6.3. Dependence of the ratio of the intensities of the reflected and incident radiation on the angle of incidence φ_0 at $\varphi_0 > (\varphi_0)_{max}$

$\varphi_0/(\varphi_0)_{max}$	1	1.5	2	5
I/I_0	1	0.02	0.005	0.0001

$$\sqrt{\frac{I}{I_0}} = \frac{2\varphi_0^2 - (\varphi_0)_{max}^2}{(\varphi_0)_{max}^2} - 2i\frac{\varphi_0\sqrt{(\varphi_0)_{max}^2 - \varphi_0^2}}{(\varphi_0)_{max}^2} = \rho\cos\alpha + i\rho\sin\alpha.$$

In this equation, the interesting feature is the value of the modulus ρ, which determines the value of the complex expression (the value α determines the phase of the reflected wave). The values of ρ and α may be determined by equating the real and imaginary parts of the resultant equation followed by the solution of the system of two equations. After simple calculations:

$$\rho = 1 \quad \text{and} \quad \text{tg}\alpha = \frac{2\varphi_0\sqrt{(\varphi_0)_{max}^2 - \varphi_0^2}}{(\varphi_0)_{max}^2 - 2\varphi_0^2},$$

which means that $\frac{I}{I_0} = \rho^2 = 1$. Thus, for transparent media at the angles $\varphi_0 < (\varphi_0)_{max}$, as also at $\varphi_0 = (\varphi_0)_{max}$, the intensity of reflected radiation is equal to the intensity of the incident radiation.

Attention will be given to the case in which the absorption of radiation in the reflecting medium cannot be ignored. If $\varphi_0 = 0$, equation (6.28) shows that:

$$\frac{I}{I_0} = \frac{a^2 + b^2}{a^2 + b^2} = 1. \tag{6.31}$$

If $\varphi_0 \ll (\varphi_0)_{max}$, the reflectivity I/I_0 may be calculated from the equation

$$\frac{I}{I_0} \approx 1 - \sqrt{\frac{2\beta}{\delta^2}}\varphi_0. \tag{6.32}$$

If $\varphi_0 = (\varphi_0)_{max}$, from equation (6.29):

$$a = b = \sqrt{\beta}.$$

Substituting the values of a and b into equation (6.28) gives:

$$\frac{I}{I_0} = \frac{\left((\varphi_0)_{max} - \sqrt{\beta}\right)^2 + \beta}{\left((\varphi_0)_{max} + \sqrt{\beta}\right)^2 + \beta} = \frac{(\varphi_0)_{max}^2 + 2\beta - 2(\varphi_0)_{max}\sqrt{\beta}}{(\varphi_0)_{max}^2 + 2\beta + 2(\varphi_0)_{max}\sqrt{\beta}}.$$

Taking into account that $(\varphi_0)_{max} = \sqrt{2\delta}$, this relationship is converted to the following dependence:

$$\frac{I}{I_0} = \frac{\delta + \beta - \sqrt{2\beta\delta}}{\delta + \beta + \sqrt{2\beta\delta}} < 1. \tag{6.33}$$

The determined ratio of the intensities is always lower than unity and decreases with the increase of β, which characterises the absorption properties of the reflecting medium.

If $\varphi_0 \gg (\varphi_0)_{max}$, equations (6.28) and (6.29) show that the investigated reflectivity I/I_0 asymptotically tends to φ_0 with the increase of the angle φ_0. In this case, the approximate value of this quantity is determined by the equation:

$$\frac{I}{I_0} \approx \frac{\delta^2}{4\varphi_0^4}, \tag{6.34}$$

i.e. decreases with the increase of φ_0 in proportion to the fourth power of the angle.

These results show that the concept of the limiting angle $(\varphi_0)_{max}$ has any meaning only in the case of reflecting media, transparent to x-ray radiation. In reflection from absorbing media, the intensity of the reflected x-radiation decreases with the increase of the sliding angle φ_0 from the intensity, equal to the intensity of incident radiation for $\varphi_0 = 0$, to almost zero at $\varphi_0 \gg (\varphi_0)_{max}$. The intensity of reflected radiation decreases with the increase of the angle and the rate of this decrease depends on the absorption capacity of the reflecting medium. This is confirmed by the calculated and experimental data obtained in [7] and presented in Fig. 6.4.

Figure 6.4*a* shows that at $\beta/\delta = 2$ and at $\varphi_0 = 1.5\,(\varphi_0)_{max}$ the ratio $I/I_0 \approx 0.1$. This ratio is higher than the ratio obtained at lower absorption characteristics of the reflecting medium (for example, at $\beta/\delta = 0$).

Figure 6.4*b* shows that the intensity of reflected long -wave x-ray radiation decreases quite rapidly because of the strong absorption of the radiation in the reflecting medium. At the same time, the angles at which the reflected intensity of the radiation of the investigated elements is high are relatively large.

Thus, the reflection of x-ray radiation even at the angles of incidence smaller than the critical angle $(\varphi_0)_{max}$ is accompanied by a partial decrease of radiation intensity. This is explained by absorption in the reflecting medium.

According to the reflection theory, proposed by Eikhenval'd [247], the incident wave penetrates into the reflecting medium, moves along a curvilinear path and after one-half period leaves the medium.

The intensity of the electromagnetic waves, penetrating into the medium,

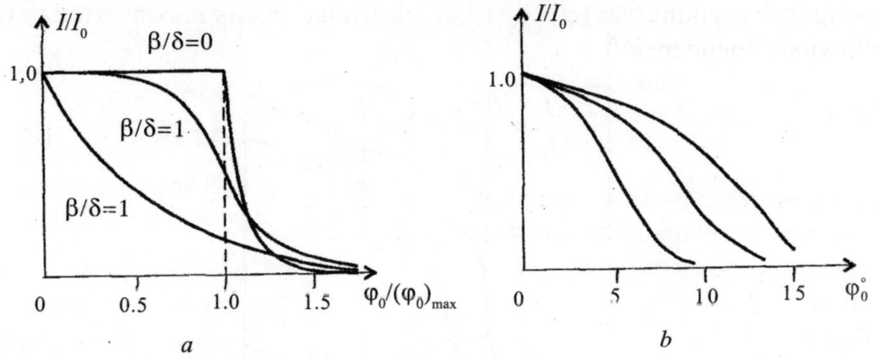

Fig. 6.4. Dependence of the ratio of the intensities of the reflected and incident x-ray radiation on the angle of incidence of radiation: (a) calculation for different absorption characteristics of the reflecting medium; (b) the experimental results for the longwave x-ray radiation of elements with small atomic numbers ($\lambda_{\text{Be K}\alpha}$=114 Å, $\lambda_{\text{B K}\alpha}$=67 Å, $\lambda_{\text{C K}\alpha}$=45 Å). Reflection from the titanium surface.

decreases with the increase of the depth in accordance with the exponential law. The depth of penetration of the radiation may be determined by the value x_e at which the intensity of radiation decreases e times [248, 249]:

$$x_e = \frac{\lambda}{2\pi\sqrt{2}\sqrt{\sqrt{(\vartheta^2-1)^2+4\beta^2}-(\vartheta^2-1)}}, \tag{6.35}$$

where $\vartheta = \varphi_0/(\varphi_0)_{max}$, λ is the wavelength of incident radiation.

At the angles considerably smaller than the critical angle, $\varphi_0 \ll (\varphi_0)_{max}$, the depth x_e may be calculated using the approximate equation (see [246]):

$$x_e \approx \frac{\lambda}{4\pi}\frac{1}{\sqrt{2\delta}}. \tag{6.36}$$

Since according to the equation (6.13) the decrement δ is proportional to the square of the wavelength, then $\lambda/\sqrt{2\delta}$ = const and the depth of penetration x_e in these cases is independent of the photon energy of incident radiation.

If the angle of incidence is equal to the critical angle, $\varphi_0 = (\varphi_0)_{max}$, then $\vartheta = 1$ and from (6.35):

$$x_e = \frac{\lambda}{4\pi}\frac{1}{\sqrt{\beta}} = \frac{1}{2}\sqrt{\frac{\lambda}{\pi\tau}}, \tag{6.37}$$

where τ is the linear coefficient of absorption of this radiation in the reflecting medium.

At larger angles, $\varphi_0 \gg (\varphi_0)_{max}$, the depth of penetration strongly depends on the absorption properties of the reflecting medium as indicated by the approximate expression, presented in [246]:

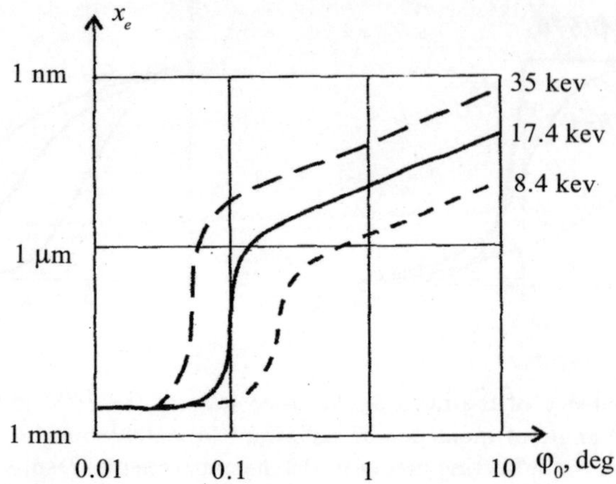

Fig. 6.5. Dependence of the depth of penetration x_e of x-ray radiation into a silicon medium on the sliding angle φ_0 for three photon energies of this radiation.

$$x_e \approx \frac{\lambda}{4\pi}\frac{\varphi_0}{\beta}. \tag{6.38}$$

Figure 6.5 shows the dependence of the depth of penetration of x-ray radiation with different photon energy on the angle of incidence of this radiation on the surface of silicon.

Figure 6.5 shows that there is a range of small angles φ_0 in which the depth of penetration of x-ray radiation into the silicon medium does not depend on the photon energy of this radiation. However, in accordance with the equation (6.37), at $\varphi_0 = (\varphi_0)_{max}$ the depth of penetration depends on τ and, consequently, is determined by the material of the reflecting medium. The minimum depths of penetration of radiation into different media, calculated using the equation (6.37), are presented in Table 6.4. The table also gives the depth of penetration into the investigated media of Mo Kα-radiation at the critical angle $(\varphi_0)_{max}$, calculated using equation (6.37).

For the absorbing media at the incidence angles higher than the critical angle $(\varphi_0)_{max}$ the processes of refraction take place simultaneously with the processes of complete external reflection of incident radiation. If the refracted beam is reflected by another surface parallel to the first surface, then there may be interference between the refracted and reflected radiations because the wavelengths of these radiations are identical.

6.4. Interference of x-ray radiation

The interference phenomenon results from the superposition of two or more

Table 6.4. The depth of penetration of radiation into different media at small angles $\varphi_0 \ll (\varphi_0)_{max}$ and Mo K-radiation at the angles $\varphi_0 = (\varphi_0)_{max}$

Medium	x_e at $\varphi_0 \ll (\varphi_0)_{max}$ (nm)	x_{efdo} for Mo Kα–radiation at $\varphi_0 = (\varphi_0)_{max}$ (nm)
Plexiglass	4.3	241
Quartz glass	3.2	83
Silicon	3.2	62
Copper	1.7	11.5
Tantalum	1.3	6.0
Gold	1.2	5.0

beams. The simplest case of interference is the summation of two waves with the same frequency and the same direction. The two-way interference of x-ray radiation may be detected in the reflection of radiation from the upper and lower surface of a thin-layer deposited on a thick substrate. In this case, the amplitude A of the resultant wave at some point is determined as follows:

$$A = \sqrt{A_1^2 + A_2^2 + 2A_1A_2\cos\varphi}, \tag{6.39}$$

where A_1 and A_2 are the amplitudes of the interfering waves and φ is the phase difference of these waves at the investigated point. If the waves are coherent, the phase difference φ at the given point is constant. Consequently, the resultant amplitude A is characterised by a constant spatial distribution with alternating minima and maxima. If $A_1 = A_2$, the maximum of the resultant amplitude is equal to the double amplitude of the combined waves, and the minimum is equal to zero. The minima are detected when the phase difference φ is equal to π, and the maxima when $\varphi = 2\pi$.

6.4.1. Reflection of x-ray radiation by the layer-substrate system and by multilayer structures

Let a monochromatic beam of x-ray radiation to fall from vacuum (or air) under a small sliding angle on a thin homogeneous layer, deposited on a flat ideally smooth thick substrate. It is assumed that the optical density of the substrate material is lower than that of the material of the layer. Consequently, at the angles close to the critical angle of complete external reflection, a process takes place whose scheme is shown in Fig. 6.6.

The difference Δ of the path of the rays I and II is determined: $\Delta = (AB+BC)-DC$. The figure shows that $AB + BC = (2d/\sin\varphi)$ and $DC = (2d/\mathrm{tg}\varphi)\cos\varphi_0$.

It must be taken into account that the wavelength λ_0 after refraction changes its value to the value λ in the layer. Therefore, to determine the difference of the paths in the wavelengths of incident radiation, the value $AB + BC$ should be multiplied by the ratio $\lambda_0/\lambda = n$, where n is the refractive index. This gives:

$$\Delta = \frac{2d}{\sin\varphi} n - \frac{2d}{\mathrm{tg}\varphi}\cos\varphi_0 .$$

However, $\cos\varphi_0 = n\cos\varphi$. Consequently $\Delta = \frac{2d}{\sin\varphi} n - \frac{2d}{\sin\varphi} n\cos^2\varphi = \frac{2d}{\sin\varphi} n\,(1-\cos^2\varphi)$, or:

$$\Delta = 2dn\sin\varphi. \tag{6.40}$$

The resultant difference of the path of the rays Δ is a function of the incidence angle φ_0:

$$\Delta = 2dn\sqrt{1-\cos^2\varphi} = 2d\sqrt{n^2-\cos^2\varphi_0} = 2d\sqrt{\sin^2\varphi_0 - 2\delta + \delta^2}.$$

Because of the small value of the angle φ_0, it may be assumed that $\sin\varphi_0 \approx \varphi_0$. Since δ is small, thc value δ^2 may be ignored. In addition, as shown previously, $2\delta = (\varphi_0)^2_{max}$. Taking this into account, from the last equation:

$$\Delta \approx 2d\sqrt{\varphi_0^2 - (\varphi_0)^2_{max}}. \tag{6.41}$$

Since according to the condition reflection takes place from the surface of the medium with the lower optical density, the oscillation phase of the electrical and magnetic vectors does not change. The interference maximum corresponds to the condition $\Delta = m\lambda_0$, where m is an integer (the difference of the path of the rays is equal to the integer of the wavelengths of the incident radiation). From this:

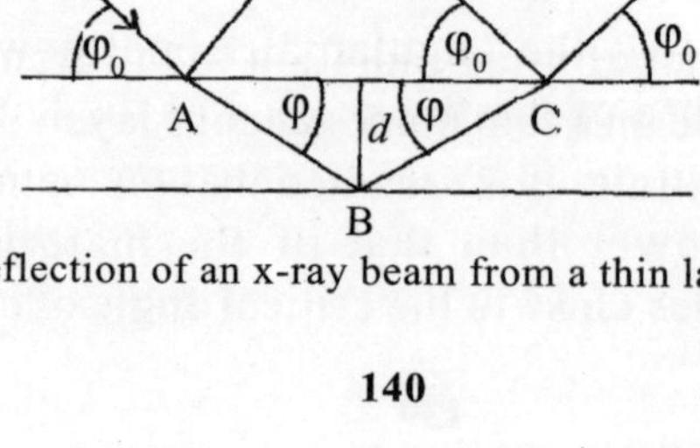

Fig. 6.6. Reflection of an x-ray beam from a thin layer on a substrate.

$$2d\sqrt{\varphi_0^2-(\varphi_0)_{max}^2}=m\lambda_0, \tag{6.42}$$

or

$$\varphi_0^2=(\varphi_0)_{max}^2+\left(\frac{m\lambda_0}{2d}\right)^2. \tag{6.42a}$$

Interference minima form when the difference of the path is equal to an odd number of the wavelengths of incident radiation. This means that:

$$2d\sqrt{\varphi_0^2-(\varphi_0)_{max}^2}=(2m+1)\frac{\lambda_0}{2}, \tag{6.43}$$

where m is an integer.

In the substrate has a higher optical density than the layer, the oscillation phase of the electrical and magnetic vectors at the layer–substrate interface changes by the value π (half the wavelength is lost). Consequently, the condition (6.42) corresponds to the interference minimum, and the condition (6.43) to the interference maximum.

Since the equations (6.42) and (6.43) have any meaning only at $\varphi_0 > (\varphi_0)_{max}$, the interference pattern is detected only in this range, i.e., outside the range of complete external reflection. Here, it should be remembered that in order to obtain the maximum value of the interference signal, the large part of incident radiation must penetrate into the layer and, at the same time, a large part of the radiation which has penetrated into the layer, should be reflected from the layer–substrate surface. In any case, the following condition should be fulfilled:

$$(\varphi_{0,l})_{max}<\varphi_0 \quad \text{and} \quad \varphi<(\varphi_{0,ls})_{max}, \tag{6.44}$$

where φ is the angle of incidence of the refracted beam on the layer-substrate surface, and the indexes l and ls refer to the surfaces of the layer and the layer–substrate, respectively. To fulfil the optimum conditions, the angle of incidence of the primary radiation on the layer with the substrate should be sufficiently small.

The equations (6.42) and (6.43) may be used to determine the angular distance between two adjacent extrema. For example, solving jointly the equations (6.42), defined for the m-th and $(m+1)$-th extremum, it is quite easy to obtain the dependence:

$$(\varphi_0)_{m+1}^2-(\varphi_0)_m^2=(2m+1)\frac{\lambda_0^2}{4d^2}. \tag{6.45}$$

Equation (6.45) shows that the angular distance between the extrema decreases with the increase of the thickness d of the layer. The thickness of the coating can be determined by measuring this distance.

In experiments, the interference pattern for the reflected radiation of the layer on the substrate was observed for the first time by Kiessig [250] and, therefore, the maxima and minima of these patterns are referred to as Kiessig patterns.

The reflectivity of the investigated system may be calculated theoretically [241] and is determined by the reflectivity of the layer and the substrate or, in the final analysis, by the sliding angle φ_0 for the given elemental composition of the layer and the substrate. The dependence of the reflectivity I/I_0 of the layer–substrate system on the sliding angle φ_0 was shown in Fig. 6.7, taken from [251].

In the case investigated in Fig. 6.7, the substrate is the medium with a higher optical density in comparison with the layer material. However, the Kiessig interference minima and maxima are also detected for the optically denser material of the layer.

The study of the layer on the substrate indicates that there may be interference of x-ray radiation in the case of multilayer structures consisting of different layers whose material is isotropic and homogeneous. The interfaces of the layers are assumed to be parallel to each other and also to the surface of the substrate. These laminated structures are usually represented by a sequence of double layers with the thickness d_1 and d_2 with alternating low and high refractive indexes n_1 and n_2. Thus, the thickness of every double layer is the sum $d = d_1+d_2$.

A special feature of the periodic multilayer structures is the reflection of x-ray radiation from them in accordance with the Wulf–Bragg law. The

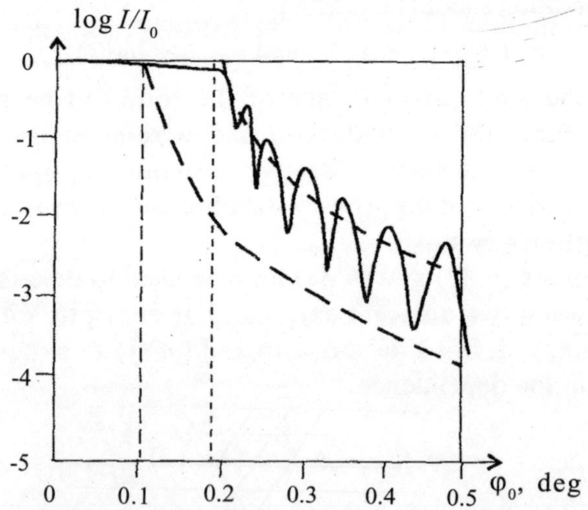

Fig. 6.7. Reflectivity I/I_0 of a thin cobalt layer (d= 30 nm) on a thick silicon substrate as a function of sliding angle φ_0. Calculations were carried out for Mo Kα-radiation (λ = 0.71 Å). The thin dashed line indicates the reflectivity of Co. The thick dashed line shows the reflectivity of Si. The thin and thick dashed vertical lines indicate the critical angles $(\varphi_0)_{max}$ for Co and Si, respectively.

reflection diagram of these structures is shown in Fig. 6.8. (Of course, the actual angles φ_0, φ_1 and φ_2 are considerably smaller than the angles indicated in the figure).

The increase of the intensity as a result of interference takes place when the difference in the path of the x-rays, reflected from two adjacent planes, is an integer of the wavelengths λ. In this condition, the reflection of radiation is determined (see [252]) by the equation:

$$2d_1 \sin\varphi_1 + 2d_2 \sin\varphi_2 = N\lambda, \tag{6.46}$$

where N is an arbitrary integer which determines the reflection order. Taking into account the small difference of the angles φ_0, φ_1 and φ_2, it is possible to write the approximate condition of reflection in the usual form:

$$2(d_1 + d_2)\sin\varphi_m \approx N\lambda, \tag{6.47}$$

where φ_m is the average angle of sliding of incident radiation for the layers d_1 and d_2.

The Wulf–Bragg law (6.47) may also be described (see the derivation of equation (6.42a)) by the dependence:

$$\varphi_m^2 \approx (\varphi_0)^2_{\max} + \left(N\frac{\lambda}{2(d_1+d_2)}\right)^2. \tag{6.48}$$

The reflections, given by the expression (6.48), are identical with the Kiessig maxima, determined by the equation (6.42a), in which N is equal to the number of the double layers of the multilayer structure. For example, if the number of the double layers is equal to 15, the Wullf–Bragg reflection maximum in the first order coincides with the 15th Kiessig maximum, in the second order with the 30th Kiessig maximum, etc. The intensity of the

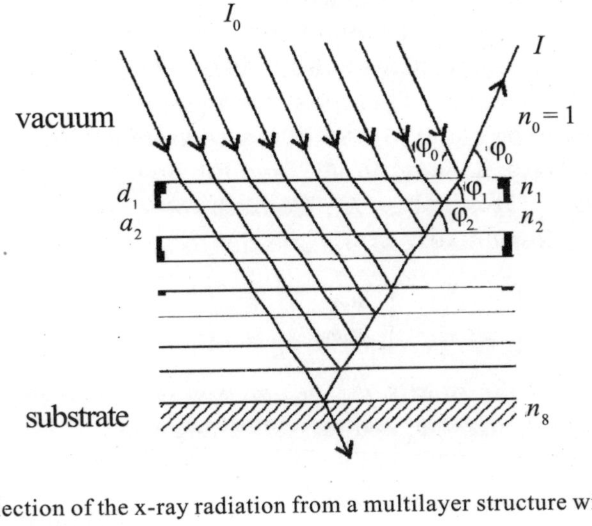

Fig. 6.8. Reflection of the x-ray radiation from a multilayer structure with the period d_1+d_2 deposited on a substrate.

Wulff–Bragg maximum is approximately an order of magnitude higher than the intensity of the corresponding Kiessig maximum and usually decreases with the increase of the reflection order.

6.4.2. Fields of standing waves

The interference, examined in the previous section, assumes the propagation of two or more beams in the same direction. However, the beams with different directions may also cause interference. If the region of superposition of the beams is large, a wave field forms with the resultant waves, which may prove to be stationary. In this case, they are referred to as standing waves with a locally stable amplitude. The local minima of the standing waves are referred to as nodes, and the local maxima as antinodes.

The simplest case of formation of standing waves is the superposition of a wide incident and a wide reflected wave. For example, in the case of x-ray radiation, standing waves may form by complete external reflection of radiation from the surface. The field of standing waves is restricted here in the longitudinal section by a triangle, formed by the surface and the fronts of the incident and reflected waves.

Like the ratios of the intensity of the incident and reflected radiation, the theoretical calculation of the intensity of standing waves may be carried out on the basis of Frenel equations within the framework of matrix formalism, proposed by Abel [253] and investigated in detail in [241, 254]. The matrix approach is extended to the cases of the reflecting surface, the layer on the substrate, and multilayer structures. The simplest case of the reflecting surface will be investigated.

In examining the ratio of the amplitude, the amplitude of a flat incident wave is normalised with respect to unity. No reflected wave is found below the surface of the material. Therefore, the hypothetical amplitude of the wave is assumed to be equal to zero. It is also assumed that the amplitude of the reflected wave is equal to E_0^{v}, and the amplitude of the refracted wave at $\varphi_0 < (\varphi_0)_{\max}$ is equal to E_s^i. Consequently, these four amplitudes are linked by the Born matrix or transfer matrix:

$$\begin{bmatrix} 1 \\ E_0^{\text{v}} \end{bmatrix} = \begin{bmatrix} m_1 m_2 \\ m_3 m_4 \end{bmatrix} \begin{bmatrix} E_s^i \\ 0 \end{bmatrix}. \tag{6.49}$$

The components of the matrix m_1, m_2, m_3 and m_4 are determined by the values of the sliding angle φ_0 and refraction angle φ_1:

$$m_1 = m_4 = \frac{\varphi_0 + \varphi_1}{2\varphi_0},$$

$$m_2 = m_3 = \frac{\varphi_0 - \varphi_1}{2\varphi_0}.$$

The amplitude of the reflected wave E_0^v is determined from (6.49) as the ratio of the components m_3 and m_1:

$$E_0^v = \frac{\varphi_0 - \varphi_1}{\varphi_0 + \varphi_1}.$$

The intensity of the standing waves is the sum of the flat waves propagating downwards and upwards, and depends on the moduli of the amplitudes of the incident and reflected waves as follows:

$$I(\varphi_0, x) = I_0\left(1 + \left|E_0^v\right|^2 + 2\left|E_0^v\right|\cos\left(\tfrac{2\pi x}{a} - \varphi(\varphi_0)\right)\right), \qquad (6.50)$$

where I_0 is the intensity of primary radiation, a is a distance between the node (or antinode) planes, x is the distance to the reflecting surface, $\varphi(\varphi_0)$ is the phase shift.

Since the amplitude of the incident wave is normalised with respect to unity, the square of the modulus $\left|E_0^v\right|^2$ determines at the same time the reflectivity $R(\varphi_0)$ for the interface of two media. Therefore, instead of (6.50):

$$I(\varphi_0, x) = I_0\left(1 + R(\varphi_0) + 2\sqrt{R(\varphi_0)}\cos\left(\tfrac{2\pi x}{a} - \varphi(\varphi_0)\right)\right). \qquad (6.51)$$

Thus, the intensity of the standing waves is determined (see [255]) by the incidence angle φ_0 and by the distance x from the reflecting surface.

The intensity of the nodes and antinodes corresponds to the maxima and minima of the equation (6.51):

$$I_{\max,\min} = I_0\left(1 + R(\varphi_0) \pm 2\sqrt{R(\varphi_0)}\right) \qquad (6.52)$$

If it is assumed that the reflectivity $R = 100\%$, the contrast between the maximum and the minimum is: $I_{\max} = 4$, $I_{\min} = 0$. This is explained by the fact that in the summation of the identical amplitudes of the incident and reflected waves, they are either doubled (in the antinodes) or decrease to zero (in the nodes). The intensities, determined as the square of the amplitude, have the values $4I_0$ and 0, respectively.

The intensity of the refracted wave at $\varphi_0 < (\varphi_0)_{\max}$ decreases exponentially with the depth x in accordance with the equation:

$$I(\varphi_0, x) = I_0 \left|E_s^i\right|^2 \exp\left\{-\tfrac{x}{x_e}\right\}, \tag{6.53}$$

where x_e is the depth of penetration, determined by the equation (6.35), and the amplitude of the refracted wave E_s^i, as indicated by the matrix equation (6.49), is equal to $\dfrac{1}{m} = \dfrac{2\varphi_0}{\varphi_0 + \varphi_1}$. The square of this amplitude determines the transmissivity of the reflecting material and may be expressed by means of reflectivity $R(\varphi_0)$. After substituting the square of the amplitude $\left|E_s^i\right|^2$ in (6.53) by reflectivity:

$$I(\varphi_0, x) = I_0 C(1 - R)\frac{\varphi_0}{x_e} \exp\left\{-\tfrac{x}{x_e}\right\}, \tag{6.54}$$

where $C = \dfrac{\lambda}{4\pi}\dfrac{1}{\beta}$, and β is the characteristic of the absorptivity of the reflecting material.

The dependence of the intensity of Mo Kα-radiation on the distance to the surface and on the depth of penetration for the angle $\varphi_0 = (\varphi_0)_{max}$ and for reflectivity $R = 81.5\%$, is shown in Fig. 6.9.

Figure 6.9 shows that in the investigated case, the first antinode coincides with the reflecting surface of silicon, and the nodes above this surface follow at a distance of a = 20 nm, and below the surface the intensity attenuates at a distance of several tens of nanometres. At the reflectivity of $R(\varphi_{0max}) = 81.5\%$, the intensity of the antinodes is 3.6 times higher than the intensity of incident radiation.

At the critical angle $(\varphi_0)_{max}$, the period of the standing waves is determined by the expression:

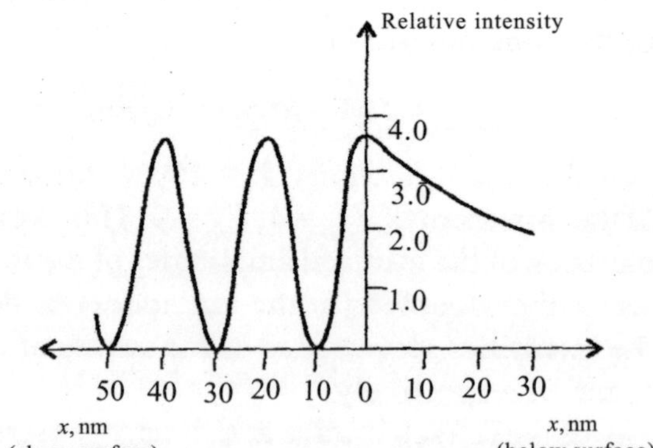

Fig. 6.9. Intensity above and below the silicon surface for Mo Kα-radiation in incidence of this radiation on the surface under the angle $(\varphi_0)_{max} = 0.1°$.

Table 6.5. Dependence of the period of the standing waves a_{crit} at the critical angle of incidence $(\varphi_0)_{max}$ on the material of the reflecting surface

Reflecting material	a_{crit}, nm
Plexiglass	26.9
Quartz glass	20.3
Silicon	20.0
Copper	10.7
Tantalum	8.3
Gold	7.7

$$a_{crit} = \frac{\lambda}{2}\frac{1}{\sqrt{2\delta}}, \tag{6.55}$$

i.e. does not depend on the wavelength of reflected radiation (since the decrement δ is proportional to the square of λ). The periods of the standing waves in this case for several reflecting surfaces are shown in Table 6.5.

At the angles smaller than the critical angle, the distance between the nodes increases, and the first antinode is displaced from the surface. The extent of penetration into the reflecting material rapidly decreases (for the angle of incidence $\varphi_0 = 0.05°$ to several nanometres).

Standing waves also form in the case of a layer deposited on the thick substrate. In the triangular area I above the surface of the layer (Fig. 6.10), the beams interfere, intersect under angle $2\varphi_0$, and in the trapezoidal area II under the angle $2\varphi_1$. The radiation penetrating into the substrate (region III) propagates with a decrease of its intensity. The interference of the two beams, reflected in the same direction from the surfaces of the layer and the substrate (region IV), was investigated in the previous section.

The theoretical relationships for the intensity of the standing waves, and also the reflected and refracted radiation are obtained in this case using two matrix equations of the type (6.49), written for the external surface and for the surface of the interface between the layer and the substrate.

The necessary condition for the formation of standing waves inside the layer is that the angle of incidence should be greater than the critical angle $\varphi_0 > (\varphi_0)_{max}$. At some angles of incidence between the surfaces of the layer and the substrate there will be an integer number of the one half periods of the standing waves. If this number is even, a resonance forms, corresponding to the minimum reflectivity. In the case of an odd number of

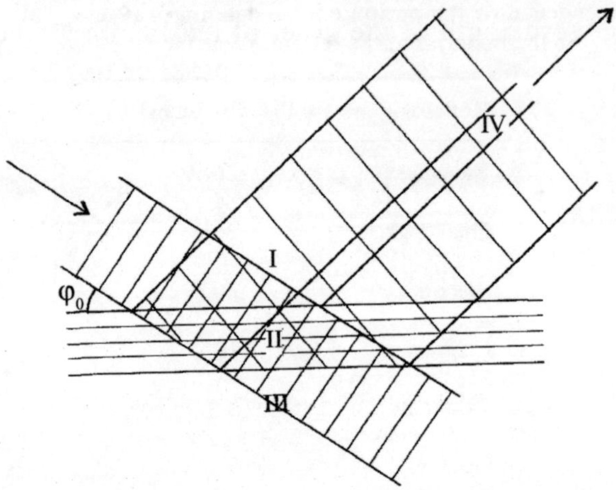

Fig. 6.10. Regions of interference above and inside the thin layer on the substrate. The standing waves form in the regions I and II.

one half cycles an anti-resonance occurs, corresponding to the maximum reflectivity. This is accompanied by a large increase of the intensity and the layer may be regarded as the x-ray waveguide [256].

Fields of standing waves are also found in multilayer structures and single crystals. In this case, the theoretical examination of the relationship of the intensities was carried out using the same matrix transfer equations (6.49).

For perfect crystals, the formation of the standing waves, based on the coherent scattering radiation by ordered atomic structures, has been investigated in [257, 258] within the framework of the dynamic theory of diffraction.

6.5. Practical application of the optical properties of x-ray radiation

The laws of refraction and reflection of x-radiation are used widely in x-ray spectrometry, diffractometry and microscopy. The application of these laws greatly improves the parameters of measuring equipment and often produces qualitatively new results for the investigated substances and materials.

Some of the achievements in this area will be examined.

6.5.1. Expanding the possibilities of x-ray spectrometry

Until recently, the expansion of x-ray radiation into a spectrum was based on the application of only natural and artificial crystals, in which the plane spacing is too small to obtain information on the spectral distribution in the area of the wavelength of the order of tens and hundreds of nanometres. The multilayer structures have made it possible to produce pseudocrystals with a large plane spacing and, at the same time, extend the application of x-ray spectrometry to the long-wave region. The currently available wave x-ray spectrometers record the irradiation of elements such as boron, carbon, nitrogen, oxygen and fluorine.

The multilayer x-ray structures (mirrors), shown in Fig. 6.8, are characterised by alternating carbon–tungsten, carbon–titanium, carbon-nickel and other periodic layers. The selection of the pair of the elements for the alternating layers is determined not only by the x-ray optical properties but also by the need for ensuring lower rates of mutual diffusion between the layers. The periods of the multilayer reflecting structures amount to 3–5 nm.

The dependence of the limiting number of the cycles N_{max} for the carbon-metal period on the wavelength of reflected radiation is shown in Fig. 6.11. The substrate for preparation of these structures is made from glass or polished quartz.

The presence of a jump in Fig. 6.11 is determined by the jump of carbon adsorption at 43.7 Å.

The theoretical expression for the ratio of the intensities of reflected radiation in the first order I_1 to incident radiation I_0 is determined by the dependence [10]:

$$\frac{I_1}{I_0} = \text{th}^2 \left\{ \frac{Nd^2}{\pi} (F_a - F_b) \right\}, \tag{6.56}$$

where N is the number of layers, d is the period, including the total thickness

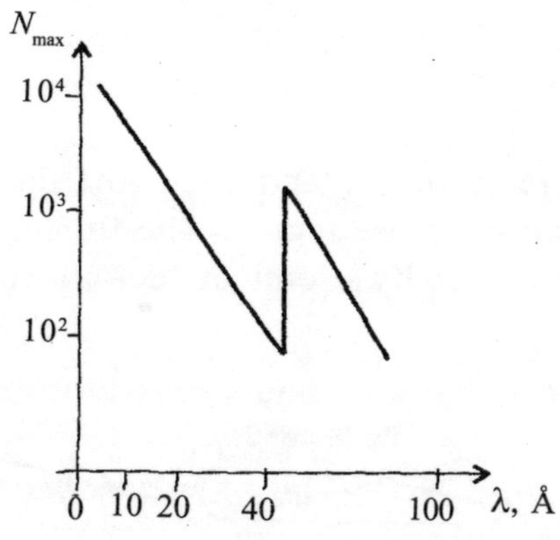

Fig. 6.11. The maximum number of carbon–metal periods of the x-ray mirror for different wavelengths of incident radiation.

of the two layers, F_a and F_b are the atomic factors of scattering for the materials a and b of these layers.

The method for preparation of multilayer coatings has been sufficiently developed. It is based on the methods of electron beam or laser sputtering with efficient control of the thickness of layers using a reflectometer. The multilayer mirrors are highly suitable from the viewpoint of crystallography because they do not have the mosaic (block) structure, typical of natural crystals. Regardless of this, the resolution capacity of the multilayer structures is greatly inferior to that of the single crystals and equals several degrees. For these mirrors, we should consider reflection bands rather than reflected wavelengths.

Another important direction in the application of the wave properties of x-ray radiation is the development of x-ray spectrometers with total external reflection [246, 259] which greatly reduce the limit of detection of elements in x-ray fluorescence analysis (the limit of detection is the minimum number of analysed matter which may be detected by the given method). If a monochromatic beam of x-ray radiation is directed onto the investigated object under the angle $\varphi_0 < (\varphi_0)_{max}$ then, as previously mentioned, this radiation penetrates into the reflecting material to a small depth of the order of tens of angstroms and excites x-ray fluorescence there. In the irradiated objects, primary radiation does not undergo any considerable scattering. Therefore, the background component of x-ray fluorescence, determined by this process, is extremely small and, consequently, the contrast of the analytical signal is high. In another variant, a small amount of substance to be analysed is deposited on the surface of the object which does not generate natural characteristic radiation in the investigated range of the spectrum. This substance fluoresces under the effect of incident monochromatic radiation and there is almost no background reflection from the substrate as a result of the effect of total external reflection. The principal diagram of this device, which may be useful for realisation of the two investigated variants, is shown in Fig. 6.12.

In Fig. 6.12 the reflectors 1 and 2 cut-off the short-wave part of inhomogeneous radiation of the source. The fluorescent radiation of the specimen is usually recorded by a semiconductor detector. The source may

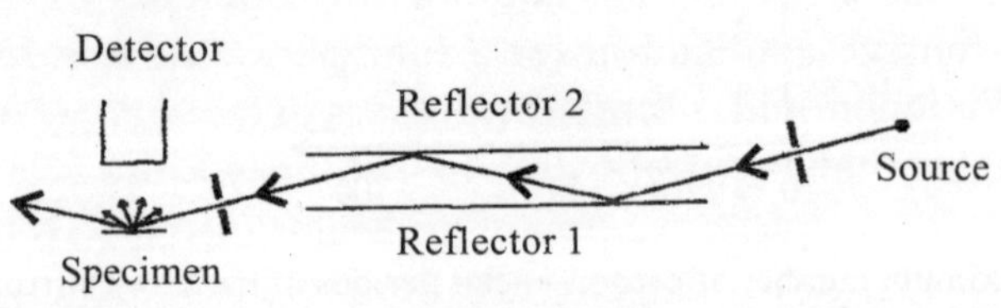

Fig. 6.12. A diagram of a fluorescent device utilising total external reflection.

have the form of an x-ray tube or synchrotron radiation. In the devices of this type there are very stringent requirements on the surface finish: surface roughness should not exceed 1 μm. The resultant limit of detection is characterised by the values of the order of 10^{-5} %. When using synchrotron radiation it is also possible to determine femtogram amounts of certain elements in the subsurface layers of the specimen [260].

The diagram shown in Fig. 6.12 was modified in [261, 262] by reducing to the maximum extent the difference between the reflectors 1 and 2 and the primary radiation source (x-ray tube) and by using a very small gap between these reflectors. Polished silicon reflectors are pressed to each other with the force of 0.1 N. Since the surfaces of the reflectors are not ideal, a gap of the order of 20–30 nm forms between them. Penetrating into this the gap, the primary radiation undergoes many times complete external reflection with the intensification of the long-wave part of the range, corresponding to the angles $(\varphi_0)_{max} > \varphi > 0$. By moving this 'slitless' collimator to the primary radiation source it is possible to increase greatly the intensity of the slightly diverging radiation beam, emitted from the gap.

6.5.2. Methods of focusing x-ray radiation

One of the aims of developing powerful wide-aperture x-ray microbeams is to increase the quality of microprobe investigations of matter. The application of an x-ray microbeam instead of the beam of charged particles in the determination of elemental composition reduces approximately 10^4 times the energy introduced into the specimen. If the intensity of the incident beam is more than 10^8 photons/μm^2·s, the detection limit decreases by an order of magnitude. The x-ray microprobe may also be used in the identification of phases and stresses with submicron resolution. The resultant resolution power and determination of strains is higher than 1 part per 10^4.

The x-ray microprobe does not usually require evacuation, and the x-ray microbeam, because of its high penetrating capacity, can be used for nondestructive analysis of deeper layers of the specimen than the charged particle probe.

Although these and other advantages of the x-ray microprobe are obvious, the low power of the x-ray sources and the low efficiency of x-ray optics restricted its extensive application for a long period of time. At present, synchrotron radiation provides for the brightness of the source 10–12 orders of magnitude higher than the brightness of x-ray tubes with a rotating anode. On the other hand, the optics of x-ray microbeams (see [132]) uses highly efficient systems with the focusing of reflected radiation, Frenel zone plates, conical capillaries and various combinations of these optical elements.

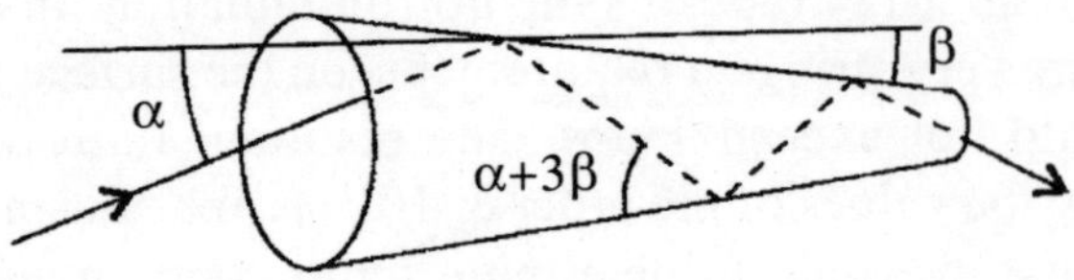

Fig. 6.13. Diagram of a linear conical capillary with several reflections of the x-ray which penetrated into the capillary.

The focusing of x-ray radiation with a conical capillary will be examined. This method produces beams (see [263]) with the size of 0.1–10 μm (the total width at half the maximum). The focusing diagram is shown in Fig. 6.13.

The x-rays undergo a complete external reflection from the smooth internal surface of the capillary at the incidence angles α smaller than the critical angle. The first reflection takes place under the angle $\alpha+\beta$, where β is the half angle of the capillary cone. The second reflection takes place under the angle $\alpha+3\beta$, and the n-th reflection under the angle $\alpha+(2n-1)\beta$. For incidence under the angle greater than the critical angle, the x-ray beam is absorbed by the surface and does not take part in the focusing process.

Maximum compression M_{max} of the beam by the conical capillary is determined as $M_{max} \geqslant (\alpha+\beta)\,(\varphi_0)_{max}$.

At the energy of the x-ray quanta of the order of 20 keV, the critical angle for glass $(\varphi_0)_{max} \approx 1.5$ mrad. Consequently, at $\alpha+\beta \approx 36$ μrad the maximum compression is of the order of 1:14.

Possibilities of capillary optics in solving the problems of x-ray analysis have been examined in a review in [264]. The radiation, reflected from the walls of the capillary, should interfere with the formation of standing waves. The wave theory of passage through the capillary of both coherent and non-coherent radiation was developed in [265, 266].

One of the shortcomings of the focusing capillaries is their small aperture. Therefore, these capillaries are promising for use as the final focusing stage of condensed systems when the comparatively smaller compression (for example 1:10) reduces the 10 μm diameter of the focus to 1 μm.

Another shortcoming of the capillaries is the problem in producing the smooth surface of the capillary orifice.

Focusing in reflection of x-ray radiation by a concave mirror includes not only the processes of complete external reflection but also the Bragg spectral optics, and also the reflection by multilayer structures. X-ray spectrum optics have been examined in detail in a textbook [267].

The diagrams of the vertical and flat linear focusing of x-ray radiation have been examined in [132, 268] and are shown in Fig. 6.14.

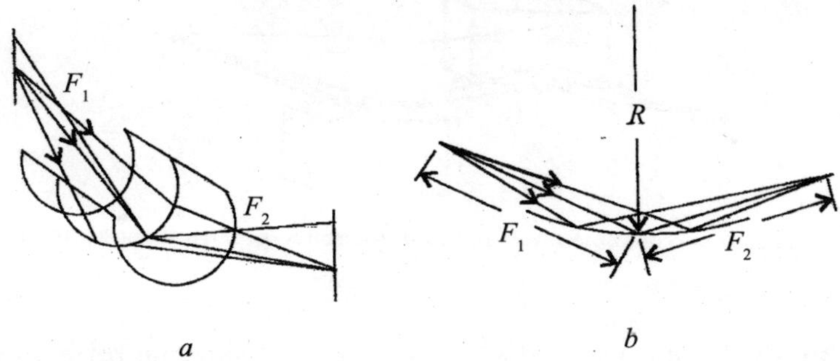

Fig. 6.14. Vertical (*a*) and flat (*b*) linear focusing of the x-ray radiation beam.

The linear schemes of focusing are used widely in x-ray spectrum apparatus with wave dispersion. In crystal-diffraction spectrometers, focusing is usually carried out in the flat variant (Johann method), vertical focusing (Hamosh method) is used less frequently. Both these methods are suitable for multilayer structures.

A point focus can be produced in total external reflection using ellipsoidal mirrors which combine the two previously examined types of focusing in a single optical element with double curvature. In principle, the ellipsoidal mirror produces the image of the source in the form of a submicron beam. However, the construction of the ellipsoidal mirrors with total external reflection is associated with serious difficulties in the preparation of highly asymmetric optics with x-ray surface finish. The mirrors of this type are available and produce the focal spot with the size of 100 μm^2 [269].

The difficulties in the production of ellipsoidal mirrors can be eliminated by the combination of meridional mirrors in the Kirkpatrick–Bayes geometry (Fig. 6.15). In the total external reflection variant, the system of the Kirkpatrick–Bayes mirrors [270], like the ellipsoidal mirror, has the advantage of letting through the wide spectral band of incident x-ray radiation.

The investigated systems usually use cylindrical or spherical mirrors so that the focal spot with a diameter of the order of 10 μm can be produced. The transition to the elliptical mirrors results in the formation of a focal spot with the diameter of the order of 1 μm. This transition has become possible after solving the problem of producing the elliptical surface with submicron roughness [271].

The Kirkpatrick–Bayes mirror system is also used in the multilayer variant. In comparison with the total external reflection variant, the angle of reflection of the multilayer mirrors is approximately five times higher. This results in difficulties in the setting of energy outside the transmission band of the multilayer structure. The reflectivity of the multilayer mirrors

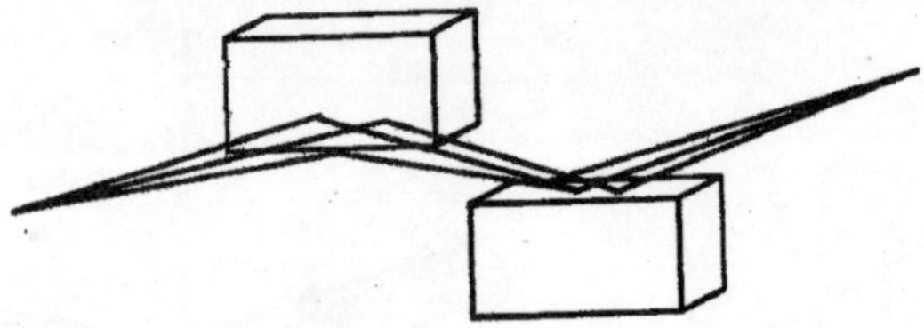

Fig. 6.15. The scheme of the path of the rays for the Kirkpatrick-Bayes focusing mirrors pair.

equals 30–60%, whereas in the case of the total external reflection mirrors it is 70–90%.

X-ray radiation may also be focused using Frenel zone plates. These plates have the form of concentric layers (zones), deposited in such a manner that the distance from the edges of the zone to the focus is $\lambda/2$, where λ is the wavelength of focusing radiation. The zone plates for the x-ray range were produced for the first time in 1988 at the Lawrence Livermore National laboratory. Using the zone plates, microbeams with the resolution of approximately 0.1 μm may be produced. The currently available technology enables the production of zone plates with the efficiency of the order of 30% and the focusing distance of ~30 cm. The focusing distance is inversely proportional to the wavelength of focusing radiation. Therefore, the resultant chromatic aberration restricts the useful aperture of the zone plate for the given spectral transmission band.

Chromatic aberration does not occur in the Bragg–Frenel optics [272, 273] which combines the reflection by the single crystal and the optics of the zone plate. Here, the Frenel contrast phase fringe is superimposed on the single crystal (or the multilayer structure) by step etching (Fig. 6.16). The phase contrast is ensured by the difference in the path length to the focus from the start and end of the step.

In practice, the formation of the zones is simplified because they rest on a thick single crystals. The narrow energy transmission band in Bragg reflection guarantees the negligibly small chromatic aberration. The size of the focus for the Bragg–Frenel spot focusing lens is $5 \times 2.5\ \mu m^2$. The Bragg-Frenel optics may be combined with the Kirkpatrick–Bayes focusing system.

The comparison of the parameters for different methods of formation of microbeams (the energy of synchrotron radiation 20 keV) is shown in Table 6.6.

Another promising direction in the focusing of the x-ray beam is the transport of radiation through bent waveguides, directed into a single spot. These waveguides are to some extent analogues of the optical fibre optics. The only difference is that vacuum is the optically densest medium for x-ray radiation. Therefore, the x-ray waveguide should be tubular.

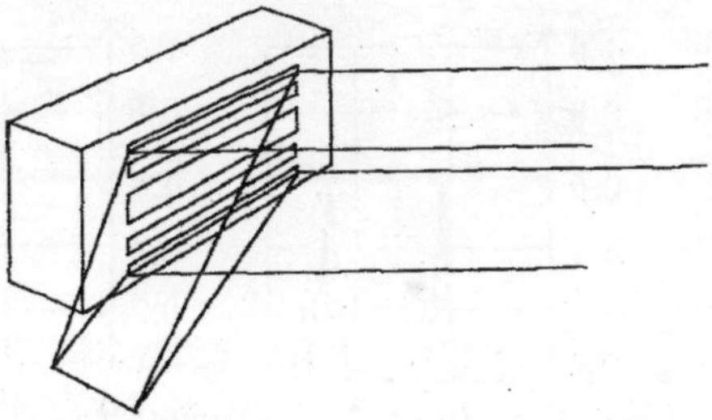

Fig 6.16. Bragg–Frenel focusing of the optical system.

Table 6.6. Comparative advantages of the methods of formation of photon microbeams with the energy of 20 keV, produced in a synchrotron

Optics	Spatial resolution (μm)	Divergence (mrad)	Flux* (photon/s)	Transmission band	Scanning
Capillary	0.1–5	~1.5	$1.6\cdot10^{10}$ monochr. $1.1\cdot10^{12}$ cont.	High E, filtered to 20 keV	Yes
Kirkpatrick–Bayes mirror	0.5–5	~ 0.5	$5.4\cdot10^{10}$ monochr. $3.6\cdot10^{12}$ cont.	Continuous to 22 keV	Yes
Kirkpatrick-Bayes multilayer	0.5-5	~ 2	$1.9\cdot10^{11}$ monochr. $1.2\cdot10^{12}$ cont.	1–5%	Difficult
Zone plate	0.2–5	~ 0.5	$5.7\cdot10^{10}$ only monochr.	$1\cdot10^{-4}$	Possible
Bragg-Frenel	0.2–5	~ 0.5	$5.7\cdot10^{10}$ only monochr..	$1\cdot10^{-4}$	Difficult

*) the flux is calculated for the focus of 1 μm^2 and 1% (continuous radiation) or 0.015% transmission band (monochromatic radiation)

The theoretical model of photon transfer of x-ray radiation through a capillary system was investigated in [274]. Reports on polycapillary x-ray lenses and their optical properties may be found in [264, 275]. The combination of the low-power pulsed sharp focus x-ray tube with the capillary half-lenses of the last generation [276] has made it possible to produce radiation pulses of the order of 10^{13} photon/s within the limits of the focal spot with the size of 20 μm.

One of the applications of capillary optics is x-ray microscopy. For example, an x-ray microscope with a parabolic capillary has been used to produce a wide-aperture focus with the size of 10 μm so that it has been

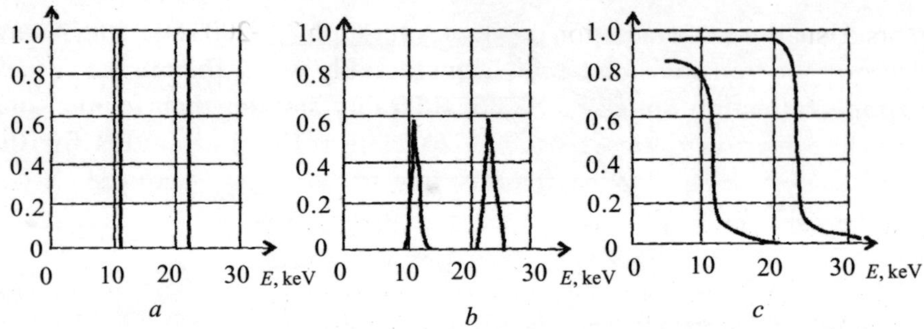

Fig. 6.17. Reflection (in the first and second order) of x-ray radiation: *a*) diffraction on the Si(111) single crystal (Wulf–Bragg diffraction), *b*) multilayer mirror, *c*) total external reflection (rhodium surface).

possible to examine microelectronic sections, crystals and a number of biological objects [277].

6.5.3. Comparison of the resolution power and reflectivity of x-ray optical elements

It is useful to compare the parameters of the processes of reflection on single crystals and multilayer mirrors, and also the identical parameters for the total external reflection effect. This comparison was carried out in [251] and is shown in Fig. 6.17.

In Bragg reflection, the reflection coefficient for perfect crystals may approach 0.8 [267]. For comparison with other reflecting elements, the reflectivity of the Bragg system is assumed to be equal to 100% (Fig. 6.17*a*). The transmission band (resolution) is determined by the number n of the planes taking part in diffraction: $dE/E \approx 1/n$, and is very narrow. Diffraction takes place at angles approximately 20 times larger than the critical angles of total external reflection. Bragg optics in especially useful for collecting greatly diverging beams of monochromatic radiation.

The total external reflection of x-ray radiation from ultrasmooth surfaces is equivalent to the total internal reflection for the visible wavelengths when the refractive index in a vacuum is lower than in a transparent medium. For mirrors with a high atomic number Z, the critical angle below which total external reflection takes place, may be described as $(\varphi_0)_{max} \approx 0.08/E$ (keV). For angles smaller than the critical angle, the reflectivity may be close to unity (Fig. 6.17*c*). The total external reflection mirrors are often used for focusing radiation with a wide transmission band.

As regards the characteristics, the multilayer synthetic mirrors occupy a middle position between the Bragg optics and the total external reflection mirrors. Usually, x-ray radiation penetrates through 20–200 reflecting layers. Therefore, the transmission band is considerably wider than in the case of the Bragg reflecting optics but is considerably smaller than for the total external reflection optics (Fig. 6.17*b*). The reflection angles for the multilayer systems are approximately five times greater than the critical angles of total external reflection.

Chapter 7

Free electrons, formed in irradiated material. Bremsstrahlung of these electrons

The ionisation of atoms by primary radiation is accompanied by the formation of a flux of photoelectrons in the specimen. The return of the ionised atoms to the initial state is accompanied by the formation of Auger electrons. In addition to this, recoil electrons are knocked out from the atom shells in non-coherent scattering of primary irradiation. All these electron fluxes which will be referred to as free electrons are sources of bremsstrahlung.

Initially, the number, spatial distribution and energy of the photoelectrons, Auger electrons and recoil electrons will be determined. This will be followed by examination of bremsstrahlung of these electrons.

7.1. Photoelectrons

The energy of the photoelectrons E_e, knocked out by a primary photon from the atom, is equal to the difference of the energy of this photon E_1 and the energy E_q, used for removing the electron from the q-shell:

$$E_e = E_1 - E_q. \tag{7.1}$$

In monochromatic primary x-ray radiation, the energy of the photoelectrons is discrete and depends on the energy E_q of bonding of the electrons in individual shells of the atom. For bremsstrahlung, the energies of the photoelectrons have the form of a continuous distribution with the limiting value of energy E_e, determined by the limiting energy of primary photons. If the energy of the primary photons approaches the binding energy of the electrons in the q-shell, the energy of the photoelectrons tends to zero.

The number of the photoelectrons n_q, formed in ionisation of the q-level, is equal to the number of the photons of x-ray radiation absorbed by this level. The ratio of the number of the photons, absorbed by the q-level of the atom, to the number of primary photons, is referred to as the partial absorption coefficient τ_q, which determines the probability of ionisation of the q-level or its ionisation cross-section (chapter 4). Thus, τ_q also characterises the probability of appearance of q-photoelectrons. It is evident that

$$n_q \propto \tau_q.$$

Or, taking into account the dependence of τ_q on the wavelength λ of ionising radiation and the atomic number Z of the irradiated material gives:

$$n_q \propto \tau_q \propto Z^4\lambda^3. \tag{7.2}$$

Thus, the number of the photoelectrons of the q-level of some element in the first approximation is proportional to the fourth degree of the atomic number of this element and the cube of the wavelength of absorbed radiation.

The distribution of photoelectrons in the exit direction will now be examined.

If the movement of the electrons in the field of the atomic nucleus is not taken into account, the direction of exit of the photoelectrons from the viewpoint of classic theory should correspond to the direction of the electrical vector of the electromagnetic field. This means that in the case of non-polarised primary radiation, the photoelectrons should be distributed in the plane normal to the direction of propagation of x-ray radiation (Fig. 7.1a).

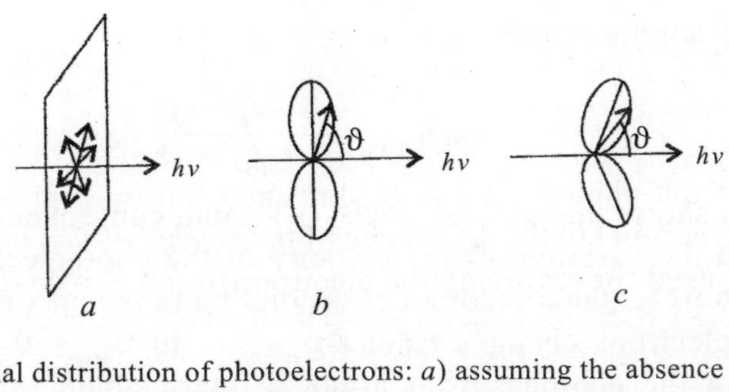

Fig. 7.1. Spatial distribution of photoelectrons: *a*) assuming the absence of the atom with the kinetic energy of the electrons, *b*) taking into account intrinsic moments of the atomic electrons at a low energy of x-ray radiation photons: *c*) taking into account intrinsic moments of the atomic electrons at a high energy of x-ray photons.

The movement of electrons in the field of the nucleus with the equally probable distribution of the directions of intrinsic moments of the electrons results in the situation in which the composition of the vectors of the external field with the fields of the moving electrons of the atom causes that the spatial distribution of the photoelectrons is no longer concentrated in the same plane. However, the distribution remains symmetric both with respect to this plane, characterising the directions of electrical vectors of nonpolarised radiation, and in relation to the direction of the primary beam. Auger and Perren showed [278] that for the case of interaction with a K-shell, the number of photoelectrons $n(\vartheta)$, which exited in the direction ϑ symmetric in relation to the primary beam is determined by the dependence:

$$n(\vartheta) \propto \sin^2 \vartheta, \tag{7.3}$$

The distribution, characterised by the expression (7.3), is shown in Fig. 7.1*b*.

In the quantum mechanics approach [279], the probability of the electron transferring from the internal level to the region of the continuous energy spectrum (proportional to the number of photoelectrons) for non-polarised x-ray radiation is determined by the following expression:

$$n(\vartheta) \propto \frac{\sin^2 \vartheta}{(1-\beta\cos\vartheta)^4}, \tag{7.4}$$

where $\beta = \upsilon/s$ is the ratio of the velocity of the photoelectrons to the velocity of light.

Taking into account equation (7.4), the direction of exit of the maximum number of the photoelectrons will be determined. The extreme value of the function (7.4) is determined from this purpose:

$$2\sin\vartheta_{max}\cos\vartheta_{max}(1-\beta\cos\vartheta_{max})^4 - 4\beta\sin^3\vartheta(1-\beta\cos\vartheta_{max})^3 = 0,$$

or

$$\beta\cos^2\vartheta_{max} + \cos\vartheta_{max} - 2\beta = 0.$$

Solving this quadratic equation gives:

$$\cos\vartheta_{max} = \frac{\sqrt{1-8\beta^2}-1}{2\beta}. \tag{7.5}$$

This equation shows that at $\beta \to 1$ $\cos\vartheta_{max} \to 1$ and, consequently, $\vartheta_{max} \to 0$.

Thus, with the variation of the velocity of the photoelectrons from a small value to $\upsilon \to c$, the direction corresponding to the maximum number of the photoelectrons changes from $\vartheta_{max} \approx \pi/2$ to $\vartheta_{max} \approx 0$. The spatial distribution of the photoelectrons at some intermediate velocity of the electrons is shown in Fig. 7.1*c*.

Experimental investigations of the spatial distribution of the photoelectrons, carried out using magnetic spectrometers and a Wilson chamber, confirmed the results of theoretical studies. The number of photoelectrons is proportional to $\sin^2\vartheta$, and the direction of exit of the largest number of the electrons shifts with the increase of the photon energy in the direction of primary beam.

7.2. Auger electrons

These electrons were discovered by Pierre Auger [280] in 1926 in examination of tracks of photoelectrons using a Wilson chamber. In addition to the tracks of the photoelectrons, Auger also detected tracks of slower electrons with the following properties:

1. The tracks appeared only simultaneously and at the same point with the primary photoelectron, but the appearance of the photoelectron was not always accompanied by the appearance of the appropriate tracks.
2. The direction of exit of these electrons did not depend on the direction of the x-ray radiation beam.
3. With the increase of the photon energy of x-ray radiation the path of the primary photoelectrons increased, and paths of the investigated electrons remained unchanged.
4. Electron energy increased with the increase of the atomic number Z of the irradiated objects.

Later, the observed electrons were referred to as the Auger electrons.

Taking into account the previously mentioned special features of the appearance of secondary electrons, the following mechanism of formation of the electrons was proposed. As a result of the photo- or some other method of ionisation, the atom is in the excited state and returns to the initial state by means of a cascade of transitions, the first of which fills the resultant vacancy on the ionised level. The energy, corresponding to the difference of the energies of the initial and final states, is lost by the atom in the form of an x-ray photon or is carried away by the electron ejected from one of its shells (Auger effect). In the second variant, the atom is in the double ionisation state. For example, if the atom looses a K-electron as a result of primary ionisation and the K-levels is filled by transition to this level of the electron from the L_2-sublevel, and the excess of energy is carried away by the Auger electron, emitted from the L_1-sublevel, then the atom no longer contains two electrons: one electron each on the L_1- and L_2-sublevels. The corresponding Auger transition of the atom is denoted by:

$$K \rightarrow L_1L_2.$$

The directions of exit of the Auger electrons are equally probable because the process of formation of these electrons is an intra-atomic process and is independent of the method of ionisation of internal shells.

The number of Auger electrons of some q-shell of the atom is proportional to the partial absorption coefficients $\tau_q(\lambda_1)$ of primary radiation for the shell multiplied by the probability of radiationless transition $(1-\omega_q)$, where ω_q is the x-ray fluorescence yield. Therefore

$$n_{oe} = (1-\omega_q)\tau_q(\lambda_1), \tag{7.6}$$

where λ_1 is the primary radiation wavelength.

The energy of the Auger electrons is determined by the energy of the levels between which transitions take place. For example, the energy of an Auger electron in the investigated case may be described by the equation

$$E_{oe} = (E_K)_Z - (E_{L2})_Z - (E_{L1})_{Z+1}. \tag{7.7}$$

The energy of the L_1-sublevel is taken for the element with the atomic number increased by unity because the absence of the electron on the L_2-sublevel increases the binding energy and makes this energy approximately equal to the energy for the element with the atomic number Z+1. More accurate calculations of the energy of the Auger electrons require introduction of corrections for the dynamic relaxation of the system of $n-1$ electrons [281, 282]. Since the method of taking these corrections into account is complicated, there is a number of semiempirical equations for calculating the energy of these electrons (see, for example, [283]).

The Auger electrons are classified on the basis of the series and groups: the series is determined by the primary vacancy (K, L, etc.), and the group by secondary vacancies (K-LL, K-LM, K-NN, etc.).

The nonrelativistic theory of the Auger effect, developed by Wentzel [284], is based on the Coulomb interaction between two electrons whose initial and final states are given by the wave functions. The relativistic examination of the Auger effect in the variant proposed by Listengarten [285] is based on the use of general equations of quantum electrodynamics for a delaying interaction of two charges regarded as a perturbation. Table 7.1 compares the experimental (for mercury, Z = 80) and calculated (for thallium, Z = 81) relative intensities of the transitions of the K-LL-group of the Auger electrons.

The table shows that there is a large difference in the probabilities of the individual Auger transitions and this assumes the existence of some selection rules established for the first time by Pincherle [287]. According to the non-relativistic theory, the relative probabilities of the Auger transitions are independent of the atomic number Z which is confirmed in the first approximation by the experimental data.

The Koster–Kronig transition for the L-, M-, etc, level, examined in

Table 7.1. Relative intensity of the Auger transitions of the K-LL-group for mercury and thallium

Transitions	$K\text{-}L_IL_I$	$K\text{-}L_IL_{II}$	$K\text{-}L_IL_{III}$	$K\text{-}L_{II}L_{II}$	$K\text{-}L_{II}L_{III}$	$K\text{-}L_{III}L_{III}$
Experimental (Z=80)*	1	1.6	0.9	0.15	1.5	0.7
Calculated (Z=81)**	1	1.70	0.91	0.06	1.31	0.55

*) average experimental data from [286], **-calculated data from [285].

chapter I, represent a variety of the Auger transitions. The difference is that the transition leading to the ejection of an electron from the outer shell, takes place between two sublevels of the inner shell. The Koster–Kronig transitions are more likely to take place than the Auger transitions and are especially intensive in the outer shells of the atom.

The atom, transferred to the double ionisation condition as a result of the primary Auger effect, reduces its energy by further transitions of the electrons from more remote shells in relation to the nucleus to the resultant vacancy. The probability of Auger transitions for the shells situated at large distances from the nucleus is higher than the probability of the radiation transitions (in the sections on the yield of x-ray fluorescence). The cascade of the Auger transitions, detected for the elements with high atomic numbers, is accompanied by the formation of a large number of vacancies on the outer shells, i.e., by the formation of the multi-charged ions [288]. For example, in the spectrum of multi-charged krypton ions, formed after the ionisation of the K-level, a large fraction is represented by atoms ionised 10–11 times. The formation of the multi-charged ions is one of the main reasons for the breakdown of complex molecules and the appearance of defects in crystals in irradiation with the flux of x-ray photons.

7.3. Recoil electrons (Compton electrons)

The kinetic energy of the recording electrons, formed in non-coherent scattering of the x-ray electrons, depends on the exit angle φ [2, 289]. In accordance with the Compton theory, the electron acquires the energy E_{kin}, equal to the energy lost by the photon during its non-coherent scattering, i.e., in accordance with (5.10):

$$E_{\text{kin}} = \Delta E = \frac{\dfrac{E^2}{m_0c^2}(1-\cos\vartheta)}{1+\dfrac{E}{m_0c^2}(1-\cos\vartheta)} = h\nu\frac{2\chi\sin^2\dfrac{\vartheta}{2}}{1+2\chi\sin^2\dfrac{\vartheta}{2}}, \tag{7.8}$$

where $\chi = \frac{h\nu}{m_0 c^2}$.

The relationship (7.8) expresses of the energy of the recoil electrons through the scattering angle ϑ and the energy $h\nu$ of the scattered photons. To transform the variable ϑ to the exit angle φ of the recording electrons, it is necessary to use the ratio of the angles ϑ and φ. Figure 5.5 shows that:

$$\operatorname{tg}\varphi = -\frac{\nu' \sin\vartheta}{\nu - \nu' \cos\vartheta}.$$

Or, expressing the quantity ν by ν':

$$\operatorname{tg}\varphi = -\frac{\sin\vartheta}{(1+\chi)(1-\cos\vartheta)} = \frac{\operatorname{ctg}\frac{\vartheta}{2}}{1+\chi}, \tag{7.9}$$

from which

$$\sin^2\frac{\vartheta}{2} = \frac{1}{1+(1+\chi)^2 \operatorname{tg}^2\varphi}. \tag{7.10}$$

Substituting the resultant equation for $\sin^2\frac{\vartheta}{2}$ into equation (7.8) gives the required dependence of the kinetic energy of the recoil electrons on the exit angle:

$$E_{\text{kin}} = h\nu \frac{2\chi \cos^2\varphi}{(1+\chi)^2 - \chi^2 \cos^2\varphi}. \tag{7.11}$$

The ratio between the angle ϑ of scattering of x-ray radiation and the exit angle φ of the recoil electrons, and also the dependence of the energy $h\nu'$ of the scattered photons and the kinetic energy E_{kin} of the recoil electrons on these angles, are shown in Fig. 7.2 in the form of a diagram proposed by Debye, in which the dotted circle gives the energy $E = h\nu$ of the photons of incident irradiation. The numbers 1–4 denotes some directions, corresponding to the scattering angles ϑ and φ of the incident photon and the exit angle of the recoil electrons. The angles φ calculated from equation (7.9) for the given value of angle ϑ, and the kinetic energies E_{kin} of the recoil electrons are calculated from equation (7.11).

Figure 7.2 shows that the sum of the energy of scattering radiation photons and the kinetic energy of the recoil electrons (see, for example, the sum of the energies E' and E_{kin} with the direction 2) is equal to the energy of the incident photon.

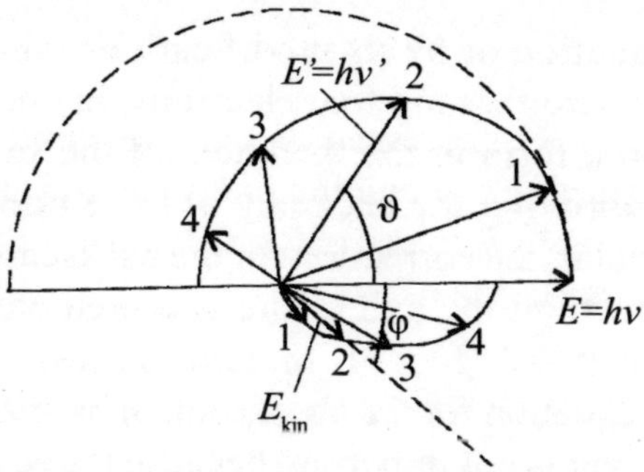

Fig. 7.2. Polar diagram of Compton scattering.

7.4. Bremsstrahlung of photoelectrons, Auger electrons and recoil electrons

The electrons formed in the irradiated material are sources of bremsstrahlung which forms during the interaction of the electrons with the nuclei of the atoms of matter. Since the free path of the electron is usually considerably shorter than the penetrating capacity of the x-ray photons, it may be assumed to a first approximation that the points of formation of these electrons and of their subsequent deceleration coincide. The spectral intensity $N(\lambda)$ of bremsstrahlung of the investigated electrons at the distance R from the specimen is described by the equation (5.45) in which the probability of scattering $d\sigma_m/d\Omega$ should be replaced by the probability $\vartheta(\lambda_1, \lambda)$ of the transformation of the primary photon with wavelength λ_1, to the photon of bremsstrahlung of the electron with wavelength λ. In addition to this, the value R^2 in this equation should be replaced by $4\pi R^2$, since the resultant bremsstrahlung is regarded as isotropic. This gives the following dependence, published in [290]:

$$N(\lambda) = \frac{sN_1}{4\pi R^2} \frac{\vartheta(\lambda_1, \lambda)}{\dfrac{\mu(\lambda_1)}{\sin\varphi} + \dfrac{\mu(\lambda)}{\sin\Psi}}, \tag{7.12}$$

Here N_1 is the number of photons of monochromatic primary radiation.

The probability $\vartheta(\lambda_1, \lambda)$ in (7.12) is the product of the probabilities $\vartheta(\lambda_1, \lambda) = \vartheta(\lambda_1, E_e)\vartheta(E_e, \lambda)$, where $\vartheta(\lambda_1, E_e)$ is the probability of transformation of the primary radiation photon with the wavelength λ_1 to a free electron with the energy E_e, $\vartheta(E_e, \lambda)$ is the probability of transformation of the free electron with the energy E_e to the bremsstrahlung photon with the wavelength λ.

The probability $\vartheta(E_e, \lambda)$ of transformation of the free electron to the bremsstrahlung photon has been studied sufficiently and can be described by the Kramers equation or by its modifications (chapter 2). It should be taken into account here that the free electrons, in contrast to the electrons of the cathode beam, form in the thickness of the specimen, and some of the corrections, improving the accuracy of the Kramers equation, are not necessary. In particular, the correction for the backscattering of the electrons which takes into account the exit of the scattered electrons to the surface of the specimen, may be ignored. In addition, the correction $f(\chi)$ in the modified Kramers equation for the absorption of bremsstrahlung of electrons in the subsurface layer is not important because the equation (7.12) contains the quantity $\mu(\lambda_1)/(\sin\varphi+\mu\,(\lambda)/\sin\psi$ which takes into account the absorption of the resultant radiation in the considerably thicker layer of the irradiated objects.

Thus, the spectral intensity of bremsstrahlung of the free electrons is determined by the Kramers equation:

$$N(E_e, E) = \text{const} \cdot Z\,(E_e - E), \tag{7.13}$$

where Z is the atomic number of the element, E is the energy of the photon in the range ΔE of the bremsstrahlung spectrum.

Then for probability $\vartheta(E_e, \lambda)$ as a function of wavelength gives:

$$\vartheta(E_e,\lambda) = \frac{13.65\cdot 10^{-6}}{L} Z \frac{(\lambda-\lambda_e)}{\lambda_e\lambda^2} \quad \text{(photon/(electron}\cdot\text{s}\cdot\text{nm)),} \tag{7.14}$$

where $\lambda_e = hc/E_e$ determines the shortwave boundary of bremsstrahlung of the electron with the energy E_e.

If the probability $\vartheta(E_e, \lambda)$ of transformation of the energy of the free electron to the energy of the bremsstrahlung photon does not depend on the method of formation of the electron, the probability $\vartheta(\lambda_1, E_c)$ of transformation of the primary photon to the free electron completely differs for each of the investigated processes. Therefore, this probabilities will be determined separately.

The probability $\vartheta_{fc}(\lambda_1, E_e)$ of the appearance of *photoelectrons* of the q-shell is determined by their number n_q, related to the number N_1 of the photons of the primary beam or, which is the same, by the number of the q-photoelectrons per single primary photon. Therefore, taking into account the unambiguous relationship (7.2) of the number of photoelectrons with the partial absorption coefficient, we obtain:

$$\vartheta_{fe}(\lambda_1, E_e) = \tau_q(\lambda_1). \tag{7.15}$$

The probability of appearance of the photoelectrons increases if the ionisation of several q-shells by primary radiation is taken into account. Summing up the photoelectrons of individual shells gives:

$$\vartheta_{fe}(\lambda_1, E_e) = \sum_q \tau_q(\lambda_1). \tag{7.16}$$

If the irradiated specimen contains several elements, each element will emit its own photoelectrons with different energy. Consequently, for a multicomponent specimen:

$$\vartheta_{fe}(\lambda_1, E_e) = \sum_i C_i \sum_q \tau_{iq}(\lambda_1). \tag{7.17}$$

where C_i is the content of the i-th element in the irradiated specimen.

Taking into account the expressions (7.17) and (7.14), the following equation is obtained for the probability $\vartheta_{fc}(\lambda_1, \lambda)$ of transformation of the primary photon to the bremsstrahlung photon:

$$\vartheta_{fe}(\lambda_1, \lambda) = \vartheta_{fe}(\lambda_1, E_e)\vartheta(E_e, \lambda) =$$

$$= 13.65 \cdot 10^{-6} \frac{Z_{eff}}{\lambda^2} \sum_i C_i \sum_q \frac{\tau_{i_q}(\lambda_1)}{L_{i_q}} \frac{(\lambda - \lambda_{i_q})}{\lambda_{i_q}}. \tag{7.18}$$

where $Z_{eff} = \sum_i C_i Z_i$ is the effective atomic number of the irradiated specimen, and λ_{iq} is the wavelength corresponding to the kinetic energy of the q-electrons of the i-th element.

The probability of formation of *an Auger electron* of the q-series with energy E_e is determined by the probability product $\tau_q(\lambda_1)$ of photoionisation of the q-shell by the probability of radiation-less transition $(1-\omega_q)$ and by the probability K_{qe} of the appearance of an electron with the energy E_e in the q-series of Auger transitions:

$$\vartheta_{oe}(\lambda_1, E_e) = \tau_q(\lambda_1)(1-\omega_q)K_{qe}. \tag{7.19}$$

For a specific Auger transition, the electron energy may be determined from the equation (7.7). In the determination of the ratio of the probabilities K_e of these transitions, the difference in the energy of the Auger electrons may be ignored (within the framework of the solved problem of search for the spectral distribution of bremsstrahlung of these electrons), if the final vacancies are situated within the limits of the same shell. However, this difference must be taken into account when the final vacancies are distributed over different shells. For example, in the K-series of Auger transitions, the difference in the electron energy in K-LL and K-LM transitions is important. To a first approximation, it may be assumed that the K-LL-transitions equal approximately 70% of the entire K-series of Auger transitions.

Restricting considerations to the range of the elements with low values of Z, it may be assumed that the bremsstrahlung is generated only by the

Auger K-electrons because these electrons of other series have a comparatively low energy. Consequently, (7.19) may be written in a different form:

$$\vartheta_{oe}(\lambda_1, E_e) = \tau_K(\lambda_1)(1-\omega_K)\sum_e K_{Ke}. \qquad (7.20)$$

To determine the probability $\vartheta_{oc}(\lambda_1, \lambda)$ of the transformation of the primary photon with the wavelength λ_1 to a bremsstrahlung photon, the equation (7.20) should be multiplied by the probability of emission of bremsstrahlung with the wavelength λ by an Auger electron given by the equation (7.14). This gives:

$$\vartheta_{oe}(\lambda_1, \lambda) = 13.65\cdot 10^{-6} Z(1-\omega_K)\tau_K(\lambda_1)\frac{1}{\lambda^2}\sum_e \frac{1}{L_e}\frac{(\lambda-\lambda_e)}{\lambda_e}K_{Ke}, \qquad (7.21)$$

where $\lambda_c = hc / E_c$ is the shortwave boundary of bremsstrahlung of the Auger electrons with energy E_{Ke}.

If the specimen contains several elements, it is important to take into account the bremsstrahlung of Auger electrons of every element. Therefore, for a multicomponent specimen, the probability of transformation of the primary photon to the bremsstrahlung photon is determined by the summation of the equations (7.21) with respect to all i elements of such a specimen:

$$\vartheta_{oe}(\lambda_1, \lambda) = 13.65\cdot 10^{-6}\frac{Z_{eff}}{\lambda^2}\times\sum_i C_i(1-\omega_{iK})\tau_{iK}(\lambda_1)\sum_e \frac{1}{L_{i_c}}\frac{\left(\lambda-\lambda_{i_c}\right)}{\lambda_{i_c}}K_{iKe}, \qquad (7.22)$$

Here C_i is the content of the i-th element and $Z_{eff} = \sum_i C_i Z_i$ is the effective atomic number of the specimen.

The probability $\vartheta_{ke}(\lambda_1, E_e)$ of the formation of a *recoil electron (Compton electron)* is determined by the probability of non-coherent scattering of the primary photon with the wavelength λ_1 in the direction in which the electron acquires the energy E_e:

$$\vartheta_{ke}(\lambda_1, E_e) = \frac{d\sigma_a^{nc}}{d\Omega}, \qquad (7.23)$$

where the atomic differential coefficient of non-coherent scattering $d\sigma_a^{\mathrm{KNT}} / d\Omega$ is determined by the Klein–Nishima–Tamm formula multiplied by the atomic factor H of non-coherent scattering (chapter 5).

Since the recoil electrons with the same energy form a cone with the tip angle ϑ, the probability of non-coherent scattering in the direction of this cone should have the form:

$$\vartheta_{ke}(\lambda_1, E_e) = \frac{d\sigma_a^{nc}}{d\Omega}(2\pi\sin\vartheta). \qquad (7.24)$$

Taking into account (7.24) and (7.14), the following equation is obtained for the probability of transformation of the primary photon to the bremsstrahlung photon of the recoil electron (for the scattering angle ϑ):

$$\vartheta_{ke}(\lambda_1,\lambda)=\frac{13.65\cdot 10^{-6}}{L}(2\pi\sin\vartheta)Z\frac{d\sigma_a^{nc}}{d\Omega}\frac{\lambda-\lambda_e}{\lambda_e\lambda^2}, \tag{7.25}$$

where λ_e is the shortwave boundary of bremsstrahlung, corresponding to the energy of the recoil electron.

The bremsstrahlung in the wavelength range $\Delta\lambda$ is formed in deceleration of all recoil electrons with different energies. Therefore, to determine the total probability of appearance of the bremsstrahlung photons in the range $\Delta\lambda$ it is necessary to integrate equation (7.25) with respect to λ_e, linked unambiguously with the energy of the recoil electrons:

$$\vartheta_{ke}(\lambda_1,\lambda)=\frac{13.65\cdot 10^{-6}}{L\lambda^2}Z\int_{\lambda_e}^{\lambda}\frac{d\sigma_a^{nc}}{d\Omega}(2\pi\sin\vartheta)\frac{\lambda-\lambda_e}{\lambda_e}d\lambda_e. \tag{7.26}$$

For calculations using equation (7.26), it is necessary to determine the dependence of $\sin\vartheta$ and $d\sigma_a^{KNT}/d\Omega$ on the integration variable λ_e. The energy of non-coherent scattering of the photons $E–E_0$ is determined by the equation (5.12):

$$E-E_e=\frac{E}{1+\chi(1-\cos\vartheta)},$$

where $\chi = E/m_0c^2$. Consequently:

$$\sin\vartheta=\sqrt{\frac{E_e}{(E-E_e)\chi}\left(2-\frac{E_e}{(E-E_e)\chi}\right)},$$

where $E_c = hc/\lambda_c$.

The atomic coefficient of non-coherent scattering is determined by the equation (5.37):

$$\frac{d\sigma_a^{nc}}{d\Omega}=\frac{r_0^2}{2}\left(\frac{E-E_e}{E}\right)\left(\left(\frac{E}{E-E_e}\right)+\left(\frac{E-E_e}{E}\right)-\sin^2\vartheta\right)Z(1-e^{-5\upsilon}),$$

where

$$\upsilon=\frac{2.21}{\lambda}\sin\frac{\vartheta}{2}Z^{-2/3},\quad \sin\frac{\vartheta}{2}=\sqrt{\frac{1-\cos\vartheta}{2}}=\sqrt{\frac{E_e}{2(E-E_e)\chi}},$$

$$E=\frac{hc}{\lambda},\quad E_e=\frac{hc}{\lambda_e}.$$

For a multicomponent specimen it is necessary to add up the equations of the type (7.26) with respect to all elements *i* of is specimen. This gives:

$$\vartheta_{ke}(\lambda_1,\lambda) = \frac{13.65\cdot 10^{-6}}{L\lambda^2} Z_{eff} \sum_i C_i \int_{\lambda_e}^{\lambda} \left(\frac{d\sigma_a^{nc}}{d\Omega}\right)_i (2\pi \sin\vartheta) \frac{\lambda-\lambda_e}{\lambda_e} d\lambda_e, \quad (7.27)$$

The resultant equations (7.18), (7.22) and (7.27) for the probabilities of transformation of the primary photon with the given energy to the bremsstrahlung photon with the wavelength λ may be used for calculating the spectral distribution of the bremsstrahlung of photo-, Auger and Compton electrons using equation (7.12).

7.5. Comparison of spectral distributions of bremsstrahlung of photo-, Auger and Compton electrons

The bremsstrahlung of these electrons is one of the reasons for the formation of a background in the x-ray spectrum of fluorescence.

Attention will be given to the simplest case of the interaction of monochromatic radiation of the radioactive source ^{241}Am (photon energy 59.6 keV) with a magnesium matrix. The spectral distribution of bremsstrahlung was calculated using equation (7.12) where the probability $\vartheta(\lambda_1, \lambda)$ is determined by the formulas (7.18), (7.22) and (7.27) for the electrons of different types. The results of the calculations are presented in Fig. 7.3.

Figure 7.3 shows that, generally speaking, the spectral distribution corresponds to the Kramers distribution (chapter 2) with a jump at $\lambda \approx 9$ Å, determined by the absorption by the K-shell of magnesium of bremsstrahlung with the wavelength of $\lambda < 9$ Å. The bremsstrahlung of the K-photoelectrons dominates in the shortwave range. At the wavelength of $\lambda > 5$ Å, bremsstrahlung of the Compton electrons is stronger than the radiation of the K-photoelectrons. In the range $\lambda > 8$ Å, this difference is expressed by a factor of 1.5. The bremsstrahlung of the K-series of Auger elcctrons is approximately 100 times less intensive in comparison with each of the investigated components. The bremsstrahlung of L-electrons of magnesium is negligible because of the low probability of ionisation of the magnesium of the L-show by the short-wave primary radiation.

The investigated partial case does not include the entire variety of possible relationships between the bremsstrahlung spectra of the free electrons. The variation of the elemental composition of the irradiated matrix has a strong effect on this ratio.

If the primary x-ray radiation is generated by an x-ray tube, the calculations of the spectral distribution should take into account the

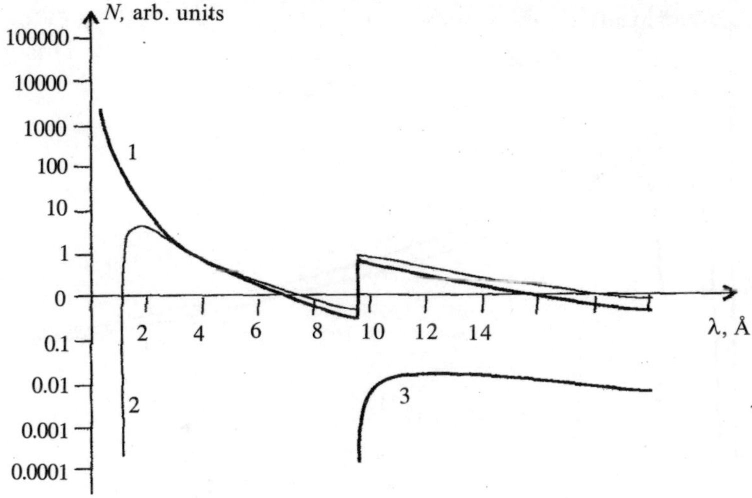

Fig. 7.3. The bremsstrahlung spectrum of K-photoelectrons (1), Compton electrons (2) and Auger electrons (3), formed in a magnesium matrix in primary radiation of ^{241}Am radioactive source with the photon energy of 59.6 keV.

polychromatic nature of this radiation. Therefore, the basic equation (7.12), valid for monochromatic radiation, should be integrated with respect to the wavelength λ_1 of the primary spectrum:

$$N(\lambda) = \frac{s}{4\pi R^2} \int_{\lambda_1} \frac{N(\lambda_1)\vartheta(\lambda_1, \lambda)\, d\lambda_1}{\dfrac{\mu(\lambda_1)}{\sin\varphi} + \dfrac{\mu(\lambda)}{\sin\psi}}, \tag{7.28}$$

where $N(\lambda_1)$ is the spectral distribution of primary x-ray radiation.

Figure 7.4 compares the bremsstrahlung of the photo- and Auger electrons for an iron matrix in primary radiation of the x-ray tube with a rhodium anode, working at a voltage of 40 kV. The thickness of the beryllium window is 50 μm.

Figure 7.4 shows that in the wavelength range smaller than the K-edge of absorption of iron (~1.9 Å) only bremsstrahlung of the K-photoelectrons is detected. The number of the L-photoelectrons is small and the bremsstrahlung of these photoelectrons may be ignored. The intensity of the bremsstrahlung of the Compton electrons in several orders of magnitude lower. In the range behind the K-edge of absorption of iron there is the bremsstrahlung spectrum of Auger electrons, whose spectral intensity at the given wavelength greater than 3 Å is higher than that for the bremsstrahlung of the photoelectrons.

The main competing factor in the formation of the background in

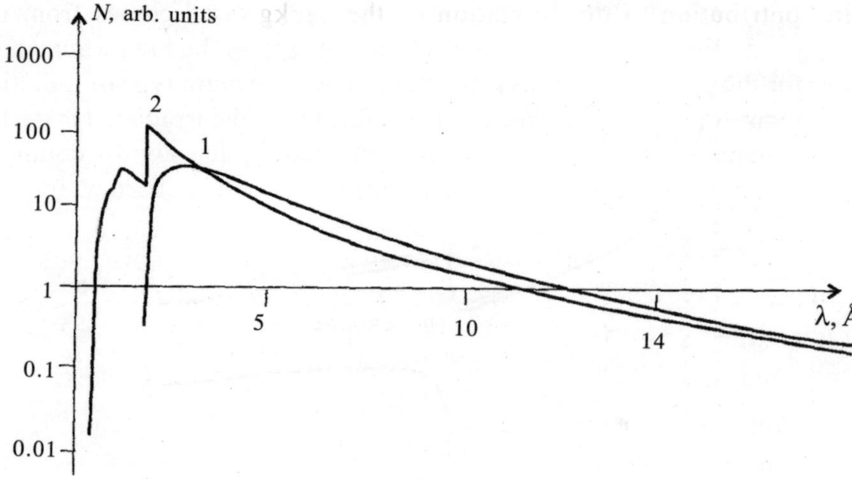

Fig. 7.4. Spectral distribution of bremsstrahlung of free electrons: 1) photoelectrons; 2) Auger electrons. Primary radiation of the x-ray tube with a rhodium cathode. The irradiated specimen was made of iron, tube potential 40 kV.

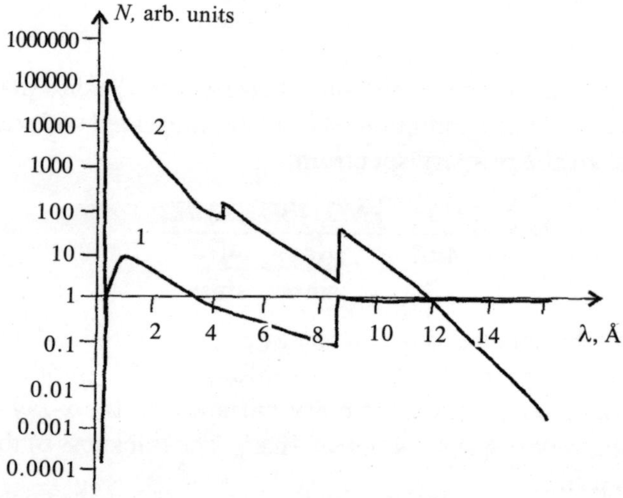

Fig. 7.5. Spectral distribution of bremsstrahlung of the free electrons, formed in an aluminium specimen (curve 1), and of the primary radiation (curve 2) scattered on the specimen for the x-ray tube with a rhodium anode (tube voltage 50 kV, the thickness of the beryllium window 50 μm).

investigations of x-ray fluorescence is the heterogeneous primary radiation, scattered on the investigated object. Figure 7.5 compares the spectral dependences of the bremsstrahlung of the x-ray tube with a rhodium anode (the thickness of the beryllium window 50 μm), scattered by an aluminium matrix, and the total bremsstrahlung of the free electrons.

Figure 7.5 shows that in the traditional range of x-ray fluorescence, the main contribution to the formation of the background comes from the primary radiation scattered on the irradiated object. In the long wavelength range (for the given conditions more than 11 Å), the main type of radiation is the bremsstrahlung of the free electrons formed in the irradiated material. The dominance of this component is explained by the almost complete absorption of the longwave primary radiation by the beryllium window of the x-ray tube.

For the multicomponent irradiated material, the spectral distribution of the bremsstrahlung of the free electrons is more complicated and depends strongly on the elemental composition of the irradiated object and on the spectral composition of primary radiation.

Chapter 8

X ray fluorescence

The photoelectric absorption of x-ray radiation, examined in chapter 4, results in the formation of vacancies in the inner shells of the ionised atoms. The filling of these vacancies with electrons away from the nucleus of the shells is accompanied, as in the ionisation with electrons and ions, by the emission of Auger electrons and of x-ray fluorescent radiation. The properties of x-ray fluorescence are of special interest for practice, because the excitation of x-ray fluorescence does not require the formation of vacuum in the region of the irradiated objects, and the process of ionisation with photons is not accompanied by the formation of bremsstrahlung which contributes greatly to the formation of the x-ray background.

8.1. Intensity of x-ray fluorescence

The theory of excitation of x-ray fluorescence in monochromatic primary radiation was developed in the 60s [291] and for inhomogeneous primary radiation was described in detail in a number of Russian monographs [26, 30, 31, 33].

The atom, ionised by a primary photon, returns to the normal state by means of cascade transitions which may include radiation transitions, i.e., the transition is accompanied by the emission of fluorescent x-ray photons. It is assumed that primary radiation is monochromatic, and the fluorescing object has the form of an infinitely elongated flat specimen (Fig. 8.1) with the content c_i of the fluorescing element i.

Consequently, the intensity of primary radiation I_x at some depth x is determined by the equation:

$$I_x = I_1 \exp\left\{-\frac{\mu(\lambda)x}{\sin\varphi}\right\}, \tag{8.1}$$

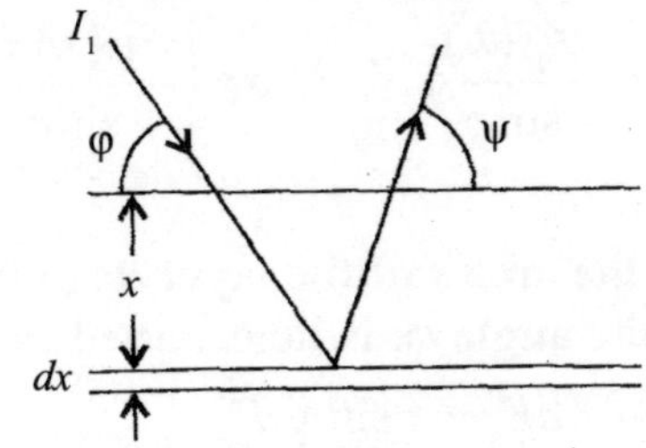

Fig. 8.1. Derivation of the equation for the intensity of x-ray fluorescence.

where μ is the linear coefficient of attenuation of primary radiation by the specimen.

The attenuation of the intensity of this radiation in the layer dx is determined by differentiation of the equation (8.1) with respect to x:

$$dI_x = -\frac{I_1\mu(\lambda)}{\sin\varphi}\exp\left\{-\frac{\mu(\lambda)x}{\sin\varphi}\right\}dx. \tag{8.2}$$

The linear attenuation coefficient $\mu(\lambda)$ is expressed through the linear coefficient of true absorption $\tau_i(\lambda)$ and scattering $\sigma_i(\lambda)$ of the individual elements i, forming the specimen. The equation $\mu_m(\lambda)=\sum_i c_i\mu_{mi}(\lambda)$ for the mass coefficient of attenuation is used for this purpose:

$$\mu(\lambda)=\mu_m(\lambda)\rho = \rho\sum_i\left(\tau_{mi}(\lambda)+\sigma_{mi}(\lambda)\right)c_i = \rho\sum_i\frac{\tau_i(\lambda)+\sigma_i(\lambda)}{\rho_i}c_i, \tag{8.3}$$

where ρ and ρ_i is the density of matter and of the i-th component, respectively, $\tau_i(\lambda)$ and $\sigma_i(\lambda)$ are the linear coefficients of true absorption and scattering for the i-th element, c_i is the mass fraction of the element i in the matter.

To examine the process of excitation of x-ray fluorescence of some element i, it is necessary to have information only on the true absorption of primary radiation by the q-shell of this element. Therefore, from equation (8.3) we use only its part associated with true absorption of radiation by the i-th element, namely $\frac{\rho}{\rho_i}\tau_{qi}(\lambda)c_i$, where $\tau_{qi}(\lambda)$ is the partial linear coefficient of true absorption of primary radiation by element i. Replacing in equation (8.2) the quantity $\mu(\lambda)$ by this relationship, we determine the attenuation of the intensity of primary radiation in the layer dx as the result of its true absorption of the atoms of the element i by the q-shell:

$$dI_x = -\frac{\tau_{qi}(\lambda)}{\sin\varphi} c_i \frac{\rho}{\rho_i} I_1 \exp\left\{-\frac{\mu(\lambda)x}{\sin\varphi}\right\} dx. \tag{8.4}$$

Energy dE_x, absorbed by the area s of the layer dx in inclined incidence of the primary beam under the angle φ, is determined as follows:

$$dE_x = s \sin\varphi \, dI_x,$$

Therefore, using the dependence (8.4) the following equation is obtained for the absorbed energy:

$$dE_{qx} = s I_1 c_i \frac{\rho}{\rho_i} \tau_{qi}(\lambda) \exp\left\{-\frac{\mu(\lambda)x}{\sin\varphi}\right\} dx. \tag{8.5}$$

The (–) in equation (8.5) is omitted, because the absorbed energy is characterised by the attenuated primary radiation with the reversed sign.

The partial coefficient of absorption $\tau_{qi}(\lambda)$ is linked with the total absorption coefficient $\tau_i(\lambda)$ by the relationship (chapter 4):

$$\tau_{qi}(\lambda) = \frac{S_q - 1}{S_q} \tau_i(\lambda).$$

Taking this relationship into account, the expression for the energy of primary radiation, absorbed by the q-level of the atoms i in the layer dx, may be presented in the form:

$$dE_{qx} = s I_1 c_i \frac{\rho}{\rho_i} \frac{S_q - 1}{S_q} \tau_i(\lambda) \exp\left\{-\frac{\mu(\lambda)x}{\sin\varphi}\right\} dx. \tag{8.6}$$

The number of photons dn_q of primary radiation, absorbed by the q-shell of the atoms i, is determined by dividing this energy by the energy of the single photon $h\nu = hc/\lambda$:

$$dn_q = s I_1 c_i \frac{\rho}{\rho_i} \frac{S_q - 1}{S_q} \tau_i(\lambda) \frac{\lambda}{hc} \exp\left\{-\frac{\mu(\lambda)x}{\sin\varphi}\right\} dx. \tag{8.7}$$

The number dn_q of the absorbed photons determines the number of the atoms of the element i with the ionised q-shell.

To determine the number of fluorescent x-ray photons of this element in the layer dx, it is necessary to take into account the fluorescence yield ω_{qi} for the q-shell. To determine the number of the photons of some j-line of the q-series, it is also necessary to take into account the probability of the transition p_{qij} with the emission of this line. Thus, the following equation is obtained for the number of x-ray photons of the line j, relating to the series q of the atoms of the element i in the layer dx:

$$dn_i = s\,I_1 c_i \frac{\rho}{\rho_i}\frac{S_q-1}{S_q}\tau_i(\lambda)\omega_{qi}p_{qij}\frac{\lambda}{hc}\exp\left\{-\frac{\mu(\lambda)x}{\sin\varphi}\right\}dx.$$

The energy of the emitted fluorescent x-ray radiation is determined by multiplying the resultant number of the fluorescent photons by their energy $h\nu_i = hc/\lambda_i$:

$$dE_i = s\,I_1 c_i \frac{\rho}{\rho_i}\frac{S_q-1}{S_q}\tau_i(\lambda)\omega_{qi}p_{qij}\frac{\lambda}{\lambda_i}\exp\left\{-\frac{\mu(\lambda)x}{\sin\varphi}\right\}dx.$$

To determine the energy of fluorescent x-radiation of the layer dx, transferred outside the limits of the specimen under the angle ψ to the surface of the specimen, it is necessary to take into account the absorption of the radiation in the layer x, i.e., to multiply the previous expression by $\exp\{\mu(\lambda_i)x/\sin\psi\}$. This gives:

$$dE_i' = s\,I_1 c_i \frac{\rho}{\rho_i}\frac{S_q-1}{S_q}\tau_i(\lambda)\omega_{qi}p_{qij}\frac{\lambda}{\lambda_i}\exp\left\{-\left(\frac{\mu(\lambda)}{\sin\varphi}+\frac{\mu(\lambda_i)}{\sin\psi}\right)x\right\}dx.$$

For the intensity of x-ray fluorescence at the distance $R \gg x$ from the probe:

$$dI_i = \frac{dE_i'}{4\pi R^2}$$

$$= \frac{s}{4\pi R^2} I_1 c_i \frac{\rho}{\rho_i}\frac{S_q-1}{S_q}\tau_i(\lambda)\omega_{qi}p_{qij}\frac{\lambda}{\lambda_i}\exp\left\{-\left(\frac{\mu(\lambda)}{\sin\varphi}+\frac{\mu(\lambda_i)}{\sin\psi}\right)x\right\}dx.$$

To determine the intensity of x-ray fluorescence for the irradiated specimen with the thickness x, the resultant expression should be integrated with respect to the depth x in the range from 0 to x. This leads to:

$$I_i = \frac{s}{4\pi R^2} I_1 c_i \frac{\rho}{\rho_i}\frac{S_q-1}{S_q}\tau_i(\lambda)\omega_{qi}p_{qij}\frac{\lambda}{\lambda_i}\int_0^x \exp\left\{-\left(\frac{\mu(\lambda)}{\sin\varphi}+\frac{\mu(\lambda_i)}{\sin\psi}\right)x\right\}dx =$$

$$= \frac{s}{4\pi R^2} I_1 c_i \frac{\rho}{\rho_i}\frac{S_q-1}{S_q}\tau_i(\lambda)\omega_{qi}p_{qij}\frac{\lambda}{\lambda_i}\frac{1-\exp\left\{-\left(\frac{\mu(\lambda)}{\sin\varphi}+\frac{\mu(\lambda_i)}{\sin\psi}\right)x\right\}}{\frac{\mu(\lambda)}{\sin\varphi}+\frac{\mu(\lambda_i)}{\sin\psi}}. \quad (8.8)$$

If the source of x-ray fluorescence is a massive specimen ($x\rightarrow\infty$), then:

$$I_i = \frac{s}{4\pi R^2} I_1 c_i \frac{\rho}{\rho_i} \frac{S_q - 1}{S_q} \omega_{qi} p_{qij} \frac{\lambda}{\lambda_i} \frac{\tau_i(\lambda)}{\dfrac{\mu(\lambda)}{\sin\varphi} + \dfrac{\mu(\lambda_i)}{\sin\psi}}. \tag{8.9}$$

In equation (8.9) we transfer from the linear coefficient of true absorption and attenuation to the mass coefficients, taking into account that: $\tau_i(\lambda) = \tau_{mi}(\lambda)\rho_i$, $\mu(\lambda) = \mu_m(\lambda)\rho$ and $\mu_m(\lambda_i) = \mu_m(\lambda_i)\rho$. This leads to:

$$I_i = \frac{s}{4\pi R^2} I_1 c_i \frac{\rho}{\rho_i} \frac{S_q - 1}{S_q} \omega_{qi} p_{qij} \frac{\lambda}{\lambda_i} \frac{\tau_{mi}(\lambda)}{\dfrac{\mu_m(\lambda)}{\sin\varphi} + \dfrac{\mu_m(\lambda_i)}{\sin\psi}}. \tag{8.10}$$

If the spectral composition of primary x-ray radiation is complicated and includes the bremsstrahlung component, then to determine the intensity of x-ray fluorescence it is necessary to integrate with respect to the wavelength in the range from the shortwave boundary of the bremsstrahlung spectrum λ_0 to the wavelength λ_q of the adsorption edge of the fluorescing element:

$$I_i^M = \frac{s}{4\pi R^2} c_i \frac{S_q - 1}{S_q} \omega_{qi} p_{qij} \frac{1}{\lambda_i} \int_{\lambda_0}^{\lambda_q} \frac{\lambda I(\lambda)\tau_{mi}(\lambda)d\lambda}{\dfrac{\mu_m(\lambda)}{\sin\varphi} + \dfrac{\mu_m(\lambda_i)}{\sin\psi}}. \tag{8.11}$$

It is necessary to determine which specimen may be regarded as massive, i.e., which layer of the specimen is saturated. The criterion of the saturated layer was determined in the monograph [26]. The ratio of the intensity I_i, determine from the equation (8.8) for the non-saturated layer, to the intensity I_i^M, calculated from (8.9) for the massive specimen, is determined by the expression:

$$\frac{I_i}{I_i^M} = 1 - \exp\left\{ -\left(\frac{\mu(\lambda)}{\sin\varphi} + \frac{\mu(\lambda_i)}{\sin\psi} \right) x \right\}. \tag{8.12}$$

It is assumed that the difference between I_i and I_i^M is 1%, i.e., $I_i / I_i^M = 0.99$.

Consequently, in (8.12)

$$1 - \exp\left\{ -\left(\frac{\mu(\lambda)}{\sin\varphi} + \frac{\mu(\lambda_i)}{\sin\psi} \right) x \right\} = 0.99.$$

Therefore

$$-\left(\frac{\mu(\lambda)}{\sin\varphi} + \frac{\mu(\lambda_i)}{\sin\psi} \right) x = \ln(0.01) = -4.6.$$

The layer with the thickness

$$x = \frac{4.6}{\frac{\mu(\lambda)}{\sin\varphi} + \frac{\mu(\lambda_i)}{\sin\psi}} \tag{8.13}$$

may be regarded as saturated with the error of up to 1%.

Equation (8.13) includes linear coefficients of attenuation of primary and fluorescent radiation. In the case of the mass attenuation coefficient, this cquation determines 1% approximation for the surface density ρx of the saturated layer (massive specimen).

The case of a very thin specimen will now be investigated [26, 292]. A criterion for the thin specimen is the following condition: the thickness of the layer x is so small that in expansion of the exponent e^{-qx} in the numerator of the equation (8.8) into a series, calculation may be restricted to the first two terms of the equation, if it is assumed that $e^{-qx} \approx 1 - qx$. In this condition, equation (8.8) gives:

$$I_i^T = \frac{s}{4\pi R^2} I_1 c_i \frac{\rho}{\rho_i} \frac{S_q - 1}{S_q} \omega_{qi} p_{qij} \frac{\lambda}{\lambda_i} \tau_i(\lambda) x, \tag{8.14}$$

Equation (8.14) shows that the thin specimen is the one for which the intensity of x-ray fluorescence does not depend on the absorption properties of the specimen, and is determined only by its density ρ, the thickness x and the content c_i of the fluorescing element.

The conditions in which the specimens may be regarded as thin will be determined. For this purpose, the ratio I_i/I_i^T of the intensity of the non-saturated layer to the intensity of the thin specimen will be examined, using the equations (8.8) and (8.14):

$$\frac{I_i}{I_i^T} = \frac{1 - \exp\left\{-\left(\frac{\mu(\lambda)}{\sin\varphi} + \frac{\mu(\lambda_i)}{\sin\psi}\right)x\right\}}{\left(\frac{\mu(\lambda)}{\sin\varphi} + \frac{\mu(\lambda_i)}{\sin\psi}\right)x}. \tag{8.15}$$

It is assumed that $\left(\frac{\mu(\lambda)}{\sin\varphi} + \frac{\mu(\lambda_i)}{\sin\psi}\right)x = y$ and also that the error of calculation of the thin layer is 1%, i.e. $I_i / I_i^T = 0.99$. Under this condition:

$$0.99 = \frac{1 - \exp\{-y\}}{y}.$$

Expanding the exponent into a series to the fourth term gives:

$$0.99 = \frac{1 - \left(1 - y + \frac{y^2}{1 \cdot 2} - \frac{y^2}{1 \cdot 2 \cdot 3}\right)}{y} = 1 - \frac{y}{2} + \frac{y^2}{6}.$$

The root of the resultant quadratic equation $y^2 - 3y + 0.06 = 0$ is $y \approx 0.02$.

Returning to the previously examined variables, for the 1% criterion of the thin layer:

$$x = \frac{0.02}{\frac{\mu(\lambda)}{\sin\varphi} + \frac{\mu(\lambda_i)}{\sin\psi}},$$

where μ and μ_i are the linear coefficients of the attenuation of primary and fluorescent radiation, respectively.

If a large number of the terms of expansion of the exponent into a series is taken into account, a more accurate value of the critical thickness x of the thin layer is obtained:

$$x = \frac{0.022}{\frac{\mu(\lambda)}{\sin\varphi} + \frac{\mu(\lambda_i)}{\sin\psi}}. \tag{8.16}$$

The expression for the surface density of the specimen $p_s = \rho x$, corresponding to the 1% approximation to the criterion of the thin layer, is determined by replacing in (8.16) the linear attenuation coefficients by mass coefficients.

8.2. Effect of the particle size on the intensity of x-ray fluorescence

The possible numerical values of the thin and saturated layers will be evaluated, using the criteria described in the previous section. It is assumed that x-ray fluorescence of calcium in its oxide CaO is excited by Cr Kα-radiation (λ = 2.29 Å). The wavelength of calcium is λ_{Ca} = 3.36 Å. The density of CaO is equal to ρ_{CaO}=1.5 g/cm^3. The mass coefficients of attenuation: $\mu_{CaO}(\lambda_{Cr})$ = 340.3 cm^2/g and $\mu_{CaO}(\lambda_{Ca})$ = 138.3 cm^2/g. It is assumed that $\varphi = \psi = 45°$.

Calculations for the investigated case using the equations (8.13) and (8.16) of the 1% approximations to the non-saturated and thin layers give:

Saturated layer: x = 45 μm;

Thin layer: x = 0.2 μm.

The size of the particles of the powder materials, prepared using standard abrasion devices, is usually in the range between the determined values,

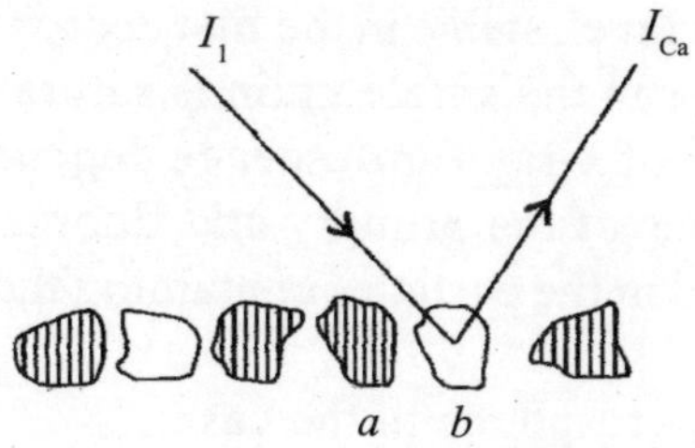

Fig. 8.2. Formation of x-ray fluorescence of a layer of the emitter with the size equal to the size of the single grain. *a*) the grains of SiO_2, *b*) grains of CaO.

i.e., corresponds to the non-saturated layer. The intensity of x-ray fluorescence in this case may greatly differ [26, 293, 294] from the value calculated from equations (8.10) and (8.11) which holds for a homogeneous fluorescing material.

It is assumed that the fluorescing material consists of the particles of CaO and SiO_2 with the same size, which is greater than the thickness of the saturated layer (Fig. 8.2). Consequently, in the first approximation, the saturated layer with the thickness of one grain will fluoresce.

If it is required to examine the intensity of x-ray fluorescence of calcium, the intensity may be approximately calculated from the equation (8.9) for the saturated layer, where the attenuation characteristics for the primary and fluorescent radiation characterise cnly the composition of the particles of CaO, containing calcium, and not the mean composition of the sample. An increase of the particle size should not influence the intensity of fluorescence. The latter claim is only approximately in agreement with the experimental data, although this may be due to the fact that the angles of incidence of primary irradiation and of the exit of fluorescent radiation for the particles depend on the shape of the particles and change with the change of the size and orientation of the particles. In the investigated case, the equation (8.10) for the intensity of x-ray fluorescence has the form:

$$I_i = \frac{k_i c_i}{\left(\frac{\mu_m(\lambda)}{\sin\varphi} + \frac{\mu_m(\lambda_i)}{\sin\psi} \right)_b}, \qquad (8.17)$$

where $\left(\frac{\mu_m(\lambda)}{\sin\varphi} + \frac{\mu_m(\lambda_i)}{\sin\psi} \right)_b$ characterises the absorption capacity of the material and relates only to the particles *b*, containing the fluorescing element. *s* in the expression for k_i denotes the area, occupied by the fluorescing particles.

Thus, with the increase of the size of the particles, the intensity of x-ray fluorescence of some element *i* changes, transferring from one stable level (the particles are so small that the fluorescing material may be regarded as homogeneous) to another level, stable in the first approximation (the particles are so large that the layer in the single grain is saturated). The increase or decrease of the intensity of x-ray fluorescence depends on the ratio of the attenuation characteristics of the primary and fluorescent radiation in the homogeneous sample and in the particles containing the fluorescing element (Fig. 8.3).

Curve 1 in Fig. 8.3 corresponds to the case:

$$\left(\frac{\mu_m(\lambda)}{\sin\varphi}+\frac{\mu_m(\lambda_i)}{\sin\psi}\right)_b < \left(\frac{\mu_m(\lambda)}{\sin\varphi}+\frac{\mu_m(\lambda_i)}{\sin\psi}\right),$$

curve 2 to the case:

$$\left(\frac{\mu_m(\lambda)}{\sin\varphi}+\frac{\mu_m(\lambda_i)}{\sin\psi}\right)_b > \left(\frac{\mu_m(\lambda)}{\sin\varphi}+\frac{\mu_m(\lambda_i)}{\sin\psi}\right).$$

In the intermediate range, in which the particle size has a strong effect on the intensity of x-ray fluorescence, it is important to evaluate the absorption properties of the inhomogeneous material. In [295, 296] it has been proposed to use the effective absorption coefficient, characterising the fine dispersion phases, whose individual grains are regarded as elementary centres of absorption. This approach has been used for describing with sufficient accuracy the absorption of x-radiation by inhomogeneous specimens with a continuous size distribution of the particles [297, 298].

Two approaches may be used for studying the intensity of x-ray fluorescence of inhomogeneous media:

a. Examination is based on expression (8.9) for a homogeneous material to which the correction γ_a for the effect of the particle size is added [26, 293]:

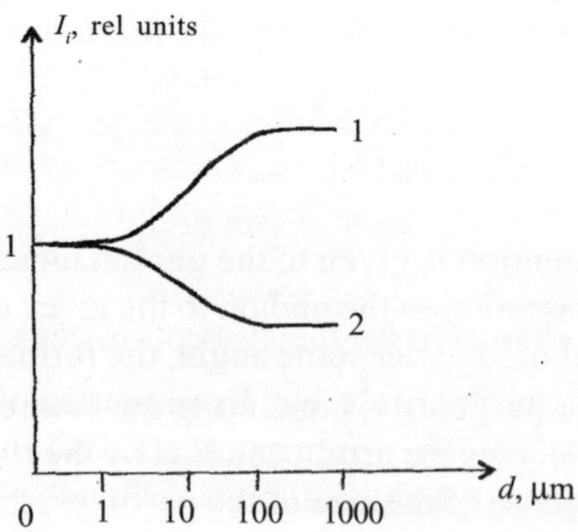

Fig. 8.3. Dependence of the intensity of x-ray fluorescence on the particle size.

$$\gamma_a = \frac{1-\exp\{-MD\}}{MD} = 1 - \frac{MD}{2!} + \frac{(MD)^2}{3!} + \ldots \tag{8.18}$$

where $M = (\mu(\lambda) + \mu(\lambda_i))_b - (\mu(\lambda) + \mu(\lambda_i))$, and D is the particle size.

Equation (8.18) shows that the correction γ_a is equal to 1 at $D \to 0$, and also for the same absorption properties of the particles, containing and not containing the fluorescing element ($M = 0$).

The introduction of this correction results in satisfactory data when the layer with the size of the single grain absorbs no more than 10% of fluorescent radiation. If this condition is not fulfilled, the accuracy of the calculations is not satisfactory.

b. Examination is based on the expression (8.17) for the intensity of x-ray fluorescence of the layer with the size of the single grain with the introduction of the correction γ_a for the deviation from the saturated layer and for the absorption in the specimen (the approach proposed in [294]):

$$\gamma_a = \frac{1-\exp\{-(\mu(\lambda)+\mu(\lambda_i))_b \rho_b D\}}{1-J_1 J_2}, \tag{8.19}$$

Here J_1 and J_2 are the exponential functions characterising the attenuation of respectively primary and fluorescent radiation in the combination of all the particles of the fluorescing object; the numerator in (8.19) determines the effect of the size and absorption properties of the grain *b*, containing the fluorescing element, on the intensity of fluorescence. This approach has been extended to the case of continuous distribution of the particle size [299].

The review, examining the effect of the heterogeneity on the intensity of x-ray fluorescence and the possible methods of taking this effect into account, has been published in [300]. The effect of heterogeneity on the intensity of fluorescence of pulpy media has been examined in [301].

In addition to the previously mentioned solutions, it is also possible to calculate the intensity of x-ray fluorescence of inhomogeneous materials by the methods of statistical evaluation (Monte Carlo method). The method (see [30, 73, 302, 303]) is based on studying possible histories of the behaviour of the x-ray photon incident on the surface of the specimen. In particular, special attention is given to the photon incident on the grain of a specific type, the penetration of the photon to the given depth, the adsorption or scattering of the photon under some angle, the formation of a fluorescent photon, its absorption in a particle, etc. Examination of the probabilities of these events for a large number of primary photons enables calculation of the intensity of fluorescent x-ray radiation of the inhomogeneous material. The results of these calculations are in good agreement with the experimental data, if all the characteristics of the inhomogeneous material are available.

Unfortunately, real powder the materials are usually sets of particles of different size and shape and different mineral and chemical composition. In addition, cavities form between the particles of the inhomogeneous specimen and the size of the cavities may equal tens of percent of the sample volume. Large particles may be coated by smaller particles. Thus, for the actual inhomogeneous specimens, calculation of the intensity of x-ray fluorescence according to the examined equations is very difficult, and the application of the Monte Carlo method find difficulty in the absence of reliable information on the state and distribution of the particles in the powder materials.

If the number of the particles, containing a fluorescing element, is small in relation to the total number of the particles, the intensity of x-ray fluorescence may change depending on the irradiated section because of the non-representative nature of the material taking part in the formation of fluorescence. This factor is especially significant if the comparatively small depth from which fluorescent radiation may exit is taken into account. The non-representative nature of the fluorescing volume is usually eliminated by increasing the irradiated area (for example, by moving the specimen in the irradiation field).

8.3. Effect of the element composition of the specimen on the intensity of x-ray fluorescence

The intensity of x-ray fluorescence will be regarded as a function of the content of the fluorescing element. For this purpose, the equation (8.10) is presented in the form:

$$I_i = \frac{k_i c_i}{\frac{\mu_m(\lambda)}{\sin\varphi} + \frac{\mu_m(\lambda_i)}{\sin\psi}}, \tag{8.20}$$

where $k_i = \frac{sI_1}{4\pi R^2}\frac{S_q - 1}{S_q}\frac{\lambda}{\lambda_i}\tau_{mi}(\lambda)\omega_{qi}p_{qj}$.

Equation (8.20) shows that the dependence $I_i = f(c_i)$ is ambiguous, because the tangent of the angle of inclination in this dependence is determined by the absorption properties of the fluorescing material, and these properties are primarily the function of the element composition (including the content of the studied element).

The dependence of the intensity of x-ray fluorescence on the content of the fluorescing element in the explicit form has been described in [268]. A sample consisting of the irradiating element i and the remaining part of the material, referred to as the filler N, is examined. It is also assumed that the elements of the filler have no absorption jumps in the range between the wavelength of primary radiation λ and the wavelength of the fluorescent line λ_i of the investigated element. Consequently, the coefficients of attenuation of primary and fluorescent radiations may be presented in the form:

$$\mu_m(\lambda) = c_i\mu_{mi}(\lambda) + c_N\mu_{mN}(\lambda),$$
$$\mu_m(\lambda_i) = c_i\mu_{mi}(\lambda_i) + c_N\mu_{mN}(\lambda_i).$$

Substituting these expressions for the absorption characteristics into the formula (8.20) and taking into account that $c_i = 1 - c_N$, and after elementary transformations:

$$I_i = k_i \frac{c_i}{\alpha_N + c_i(\alpha_i - \alpha_N\cdot)}, \tag{8.21}$$

where $\alpha_i = \frac{\mu_{mi}(\lambda)}{\sin\varphi} + \frac{\mu_{mi}(\lambda_i)}{\sin\psi}$ and $\alpha_N = \frac{\mu_{mN}(\lambda)}{\sin\varphi} + \frac{\mu_{mN}(\lambda_i)}{\sin\psi}$ are the quantities, characterising the absorption properties of the fluorescing element and the filler, respectively.

Equation (8.21) shows: if the absorption characteristics of the fluorescing element and the filler are identical (i.e., $\alpha_i = \alpha_N$), the tangent of the angle of inclination is constant and the dependence $I_i = f(c_i)$ is a straight line;

– if the absorption capacity of the fluorescing element is greater than that of the filler (i.e., $\alpha_i > \alpha_N$), the tangent of the angle of inclination decreases with increase of c_i and the dependence $I_i = f(c_i)$ is a curve convex in relation to the abscissa;

– if the absorption capacity of the fluorescing element is lower than that of the filler ($\alpha_i < \alpha_N$), the tangent of the angle of inclination increases with increasing c_i and the dependence $I_i = f(c_i)$ is a concave curve.

In all three examined cases, the dependence $I_i = f(c_i)$ has points coinciding at $c_i = 0$ and at $c_i = 100\%$, which is reflected in Fig. 8.4.

Figure 8.4 shows that the same intensity of the x-ray fluorescence I_1 corresponds to the range of the possible content of the fluorescing element from c_1 to c_2. The so-called degeneration of intensity with respect to the content of the investigated element is observed. Overcoming the resultant ambiguity in the content of the elements, corresponding to the measured intensity of the fluorescent analytical signal, is one of the main tasks of x-ray fluorescence analysis.

In (8.20), the effect of the composition of the material on the intensity of fluorescence is determined by three quantities: $\mu_m(\lambda_i)$, $\mu_m(\lambda)$ and c_i. It is interesting to combine these quantities in some factor which can be used to

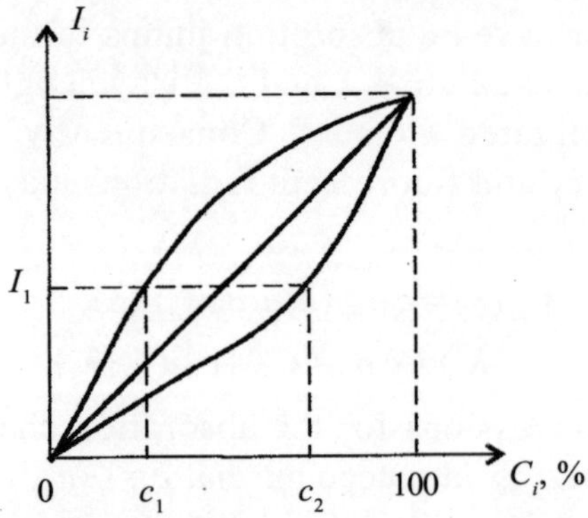

Fig. 8.4. Dependence of the intensity of x-ray fluorescence of the content of the element for different absorption properties of the specimen.

describe the set of the curves in Fig. 8.4 by a one-parameter dependence.

The ratio of the mass coefficients of attenuation by the specimen of primary $\mu_m(\lambda)$ and fluorescent $\mu_m(\lambda_i)$ radiation will be determined. For primary radiation:

$$\mu_m(\lambda) = c_i\mu_{mi}(\lambda) + \sum_{n\neq i} c_{n.}\mu_{mn}(\lambda). \tag{8.22}$$

For the fluorescing element, there is always an absorption jump S_q of this element between the wavelengths λ and λ_i. Therefore:

$$\mu_{mi}(\lambda) \approx \left(\frac{\lambda}{\lambda_i}\right)^3 S_q\mu_{mi}(\lambda_i).$$

If the remaining elements of the specimen in the range of the wavelength from λ to λ_i have no adsorption jumps, then

$$\mu_{mn}(\lambda) \approx \left(\frac{\lambda}{\lambda_i}\right)^3 \mu_{mn}(\lambda_i).$$

The resultant expression is substituted into (8.22)

$$\mu_m(\lambda) = \left(\frac{\lambda}{\lambda_i}\right)^3 \left(S_q c_i\mu_{mi}(\lambda_i) + \sum_{n\neq i} c_n\mu_{mn}(\lambda_i)\right).$$

which gives:

$$\mu_m(\lambda) = \left(\frac{\lambda}{\lambda_i}\right)^3 \left(c_i\mu_{mi}(\lambda_i)(S_q - 1) + \mu_m(\lambda_i)\right),$$

or

$$\mu_m(\lambda) = \mu_m(\lambda_i)\left(\frac{\lambda}{\lambda_i}\right)^3 \left(W_i(S_q - 1) + 1\right), \tag{8.23}$$

where

$$W_i = \frac{c_i \mu_{mi}(\lambda_i)}{\mu_m(\lambda_i)}, \tag{8.24}$$

W_i is a factor of chemical composition proposed by Blokhin and Druze [304]. It is clear that W_i changes in the range from 0 to 1, like the content c_i of the element i.

It will be shown that this factor actually takes into account all the effects of the element composition on the intensity of x-ray fluorescence. Substituting $\mu(\lambda)$, given by equation (8.23), into equation (8.20) for the intensity of x-ray fluorescence, gives:

$$I_i = \frac{k_i' c_i}{\mu_m(\lambda_i)} \frac{\tau_{mi}(\lambda)}{\left(\frac{\lambda}{\lambda_i}\right)^3 \frac{1}{\sin\varphi}(W_i(S_q - 1) + 1) + \frac{1}{\sin\psi}}, \tag{8.25}$$

where coefficient k_i' differs from k_i by the value $\tau_{mi}(\lambda)$ placed in the numerator of this equation.

The variable $\tau_{mi}(\lambda)$ can be expressed by the mass coefficient of photoelectric absorption $\tau_{mi}(\lambda_i)$ by the element i of the own radiation with the wavelength λ_i using the equation:

$$\tau_{mi}(\lambda) \approx \left(\frac{\lambda}{\lambda_i}\right)^3 S_q \tau_{mi}(\lambda_i).$$

For the majority of elements at wavelengths $\lambda_i > 0.5$ Å, the processes of scattering of the atoms are insignificant in comparison with the photoelectric adsorption of x-ray radiation by the elements. Therefore, it may be assumed that $\tau_{mi}(\lambda_i) \approx \mu_{mi}(\lambda_i)$. Consequently:

$$\tau_{mi}(\lambda) \approx \left(\frac{\lambda}{\lambda_i}\right)^3 S_q \mu_{mi}(\lambda_i).$$

Substituting the resultant expression into equation (8.25) gives the following equation for the intensity of x-ray fluorescence:

$$I_i = k_i' S_q \frac{c_i \mu_{mi}(\lambda_i)}{\mu_m(\lambda_i)} \frac{1}{\frac{1}{\sin\varphi}(W_i(S_q - 1) + 1) + \left(\frac{\lambda_i}{\lambda}\right)^3 \frac{1}{\sin\psi}}.$$

It may easily be seen that the first fraction in the resultant equation is the factor of the chemical composition W_i. Therefore, the resultant equation may be presented in the new form:

$$I_i = k_i'' \frac{W_i}{A_i W_i + 1}, \tag{8.26}$$

where

$$A_i = \frac{S_q - 1}{1 + \dfrac{\sin\varphi}{\sin\psi}\left(\dfrac{\lambda_i}{\lambda}\right)^3}$$

and $k_i'' = \dfrac{sI_1}{4\pi R^2}\dfrac{S_q - 1}{S_q}\dfrac{\lambda}{\lambda_i}\omega_q p_{qi}\dfrac{S_q \sin\varphi}{1 + \dfrac{\sin\varphi}{\sin\psi}\left(\dfrac{\lambda_i}{\lambda}\right)^3}$. The one-parameter dependence (8.26) is shown in Fig. 8.5.

Thus, in the absence of selective effects, the dependence of the intensity of x-ray fluorescence on the composition is determined by the value of the factor W_i which, as indicated by equation (8.24), depends only on the content c_i of the element i and on the mass coefficient of attenuation of fluorescent radiation by the specimen (the values of $\mu_i(\lambda_i)$ are available and tabulated in a number of tables).

8.4. Dependence of the intensity of x-ray fluorescence on primary radiation wavelength

To ensure the maximum intensity of x-ray fluorescence of some element, it

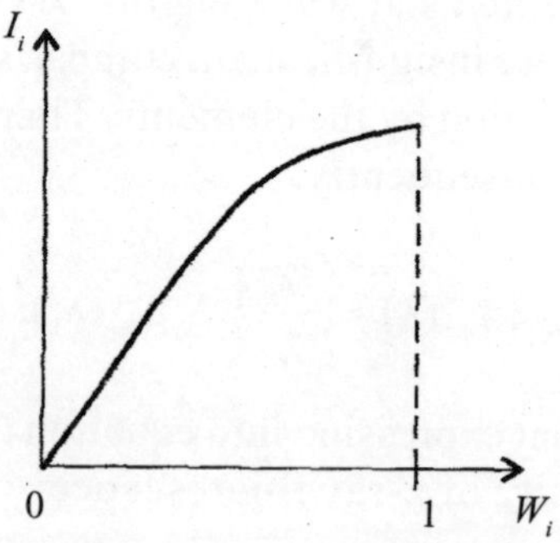

Fig. 8.5. Dependence of the intensity of x-ray fluorescence on the factor of chemical composition W_i in the absence of absorption edges (with the exception of the edge of the fluorescing element) in the range between the wavelength λ of primary radiation and the wavelength λ_i of measured fluorescence.

is important to select of the spectral composition of primary radiation. In order to determine the effect of the wavelength of primary radiation on the intensity of x-ray fluorescence, equation (8.26) will be transformed to the form in which the specific intensity (intensity related to the value of the factor of the chemical composition) is represented by the explicit function of the ratio of the wavelengths of the primary and fluorescent radiations:

$$(I_i)_{sp} = \frac{I_i}{W_i} \frac{k_i^m \left(\frac{\lambda}{\lambda_i}\right)^4 S_q}{\frac{\sin\varphi}{\sin\psi}((S_q - 1)W_i + 1\left(\frac{\lambda}{\lambda_i}\right)^3 + 1}, \tag{8.27}$$

where the coefficient

$$k_i''' = \frac{I_1 s}{4\pi R^2} \frac{S_q - 1}{S_q} \omega_q p_{qi} \sin\psi,$$

is independent of the wavelength λ.

The cases of the high (W_i ›1) and low ($W_i \to 0$) content of the fluorescing element will be examined.

At $W_i = 1$ (monoelement specimen), equation (8.27) gives:

$$(I_i)_{sp}^{W\to 1} = \frac{k_i''' \left(\frac{\lambda}{\lambda_i}\right)^4 S_q}{\frac{\sin\psi}{\sin\varphi} S_q \left(\frac{\lambda}{\lambda_i}\right)^3 + 1}. \tag{8.28}$$

The resultant dependence of the intensity of Kα-radiation of a zinc specimen on the wavelength of exciting radiation ($Z = 30$, $\lambda_i = 1.43$ Å, $\lambda_k = 1.28$ Å, $\lambda_i/\lambda_k = 0.89$, $S_k = 8$) is shown in Fig. 8.6 by the continuous curve; to simplify considerations, it is assumed that $\varphi = \psi$. This dependence is discussed in the monograph [7].

At $W \to 0$ (a low content of the element), the specific intensity is determined by the equation:

$$(I_i)_{sp}^{W\to 0} = k_i''' \frac{\left(\frac{\lambda}{\lambda_i}\right)^4 S_q}{\frac{\sin\psi}{\sin\varphi}\left(\frac{\lambda}{\lambda_i}\right)^3 + 1}. \tag{8.29}$$

The dependence of the intensity of the Zn Kα-line at a low zinc content in

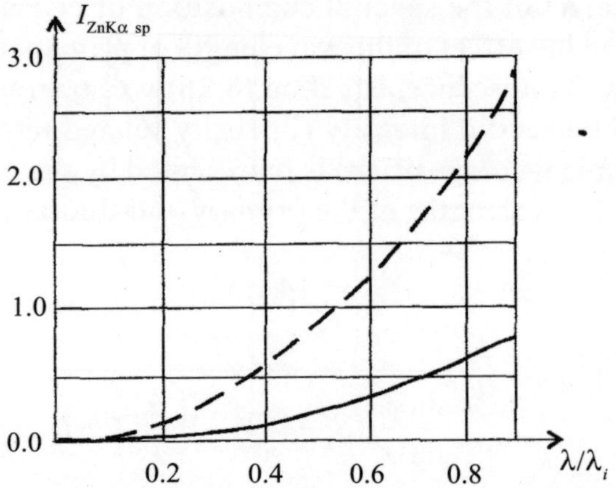

Fig. 8.6. Dependence of the specific intensity of x-ray fluorescence Zn Kα-line on the wavelengths of primary radiation. The solid line – intensity of the zinc specimen, the dotted line – specific intensity at $W_{Zn}\to 0$.

a specimen is shown in Fig. 8.6 by the broken line in the same conventional units as in the case of the zinc specimen.

Figure 8.6 shows that the dependence of the intensity of x-ray fluorescence on the wavelength is of the resonance type: as the wavelength of primary radiation moves closer to the wavelength of the absorption edge of the fluorescing element (the difference between the frequency of intrinsic oscillations of the electrons of the q-shell and the frequency of primary radiation decreases), the intensity of fluorescent radiation increases. At $\lambda > \lambda_q$, when the energy of the primary photon is insufficient for ionisation of the q-shell, the intensity of fluorescence decreases to zero.

Equations (8.28) and (8.29) show that at $\varphi = \psi$:

$$\frac{I_{sp}^{W\to 0}}{I_{sp}^{W\to 1}} = \frac{S_q\left(\frac{\lambda}{\lambda_i}\right)^3 + 1}{\left(\frac{\lambda}{\lambda_i}\right)^3 + 1}. \tag{8.30}$$

The dependence of the ratio of the specific intensities of x-ray fluorescence of the wavelength λ of exciting radiation is shown in Fig. 8.7. The ratios are calculated for the Zn Kα-line.

Figure 8.7 shows that at the short wavelength ($\lambda < 0.2\lambda_i$), the efficiency of excitation of the zinc atoms is almost independent of the zinc content of the massive specimen. The efficiency of excitation of low zinc contents increases more rapidly with the increase of the wavelength of primary radiation than in the case of high content. Thus, at the wavelengths close to

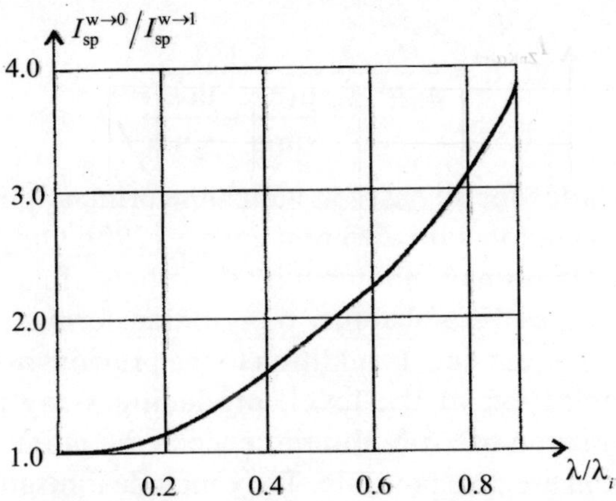

Fig. 8.7. The ratio of the specific intensities of the Zn Kα-line of x-ray fluorescence for low ($W_i \to 0$) and high ($W_i \to 1$) zinc content as a function of the wavelength of exciting radiation.

the wavelength of the K-edge of absorption of zinc, the specific intensity at a low zinc content is higher than that at $c_{Zn} \to 1$ by almost a factor of 4. This is explained by the fact that at a high content of the fluorescing element, the primary radiation with the wavelength at the absorption edge is selectively absorbed by the atoms of this element and, consequently, the depth of penetration of primary radiation into the specimen greatly decreases and the thickness of the fluorescing layer is considerably smaller than the depth from which exciting radiation may generate.

8.5. Process of excitation of x-ray fluorescence for the case in which the areas of absorption of primary radiation and formation of a fluorescent photon coincide

In particular, these processes include the previously examined photoelectric excitation in which the fluorescent photon forms in the same atom which absorbed the primary radiation photon. The value

$$\frac{S_q - 1}{S_q} \tau(\lambda) \omega_{qi} p_{qij} c_i \qquad (8.31)$$

in equation (8.8) for the intensity of x-ray fluorescence is the probability of transformation of the primary x-ray photon with the wavelength λ to the fluorescent photon with the wavelength λ_i. Denoting this probability by $\vartheta_i(\lambda)$, the following equation can be proposed for the intensity of x-ray fluorescence of the massive specimen:

$$I_i = \frac{sI}{4\pi R^2}\frac{\lambda}{\lambda_i}\frac{\vartheta_i(\lambda)}{\dfrac{\mu(\lambda)}{\sin\varphi}+\dfrac{\mu(\lambda_i)}{\sin\psi}}. \tag{8.32}$$

Equation (8.32) holds for all cases in which the primary photon transforms at some point to a photon with different energy [290]. Only the mechanism of interaction of the primary photon with the atoms of matter differs. This mechanism determines the probability $\vartheta_i(\lambda)$ of transformation of the primary photon to the fluorescent one. In addition to the previously examined direct photoelectric ionisation of the level, producing x-ray radiation, other processes of formation of x-ray fluorescence at the point of absorption of the primary photon are also possible. They include ionisation as a result of intra-atom cascade transitions and, with some approximation, the ionisation of atoms by photo- and Auger electrons, formed in the material under the effect of primary radiation. These processes will be investigated in greater detail.

8.5.1. Formation of x-ray fluorescence as a result of cascade transitions

The ionised atom returns to the initial condition by means of cascade transitions with the transfer of vacancies to the shells further away from the nucleus. The probability of appearance of the L-photon in initial ionisation of the K-shell may be determined as follows [305]. The probability of the initial ionisation of the K-shell of element *i* is determined by the equation:

$$\frac{S_K - 1}{S_K}\tau_i(\lambda)c_i,$$

where c_i is the content of the element *i* in the irradiated material.

If the total probability of radiation and radiationless intra-atomic transitions, leading to ionisation of the L-shell, is equal to P_{KL}, then the probability $\vartheta_{KL}(\lambda)$ of transformation of the primary photon to the fluorescent photon of L-radiation is described by the dependence:

$$\vartheta_{KL}(\lambda) = \frac{S_K - 1}{S_K}\tau_i(\lambda)P_{KL}\omega_L p_{Li}c_i. \tag{8.33}$$

After substituting this expression into equation (8.32), one can calculate the intensity of x-ray L-fluorescence, determined by the cascade transition. For this purpose, it is necessary to determine the probability P_{KL}.

The majority of radiation (approximately 90%) and radiationless (according to [306] more than 90%) transitions lead to ionisation of the L-shell. Therefore, the probability P_{KL} of the vacancy transferring from K- to L-shell is high:

$$P_{KL} \approx 0.9(1-\omega_K) + 0.9\omega_K = 0.9.$$

The ratio of the intensities of L-radiation, formed during direct ionisation of the L-shell by the primary photon and as a result of cascade transitions at $\lambda < \lambda_K$ will be determined. The probability of transformation of the primary to fluorescent photon in direct ionisation of the L-shell is:

$$\vartheta_L(\lambda) = \frac{S_L - 1}{S_K S_L}\tau(\lambda)\omega_L p_{Li} c_i,$$

Consequently, the ratio of the investigated intensities is determined by the equation:

$$q = \frac{I_i^{KL}}{I_i^L} = \frac{\vartheta_{KL}(\lambda)}{\vartheta_L(\lambda)} = \frac{(S_K - 1)S_L}{(S_L - 1)} P_{KL}$$

and the contribution of the cascade transitions to the excitation of L-fluorescence at $\lambda < \lambda_K$ is:

$$\frac{I_i^{KL}}{I_i} = \frac{q}{1+q}.$$

Calculations of the magnitude of the contribution shows that at $\lambda < \lambda_K$ approximately 60–70% of the total intensity of L-radiation is determined by the ionisation of the L-shell as a result of cascade transitions.

In addition to the cascade transitions, at a high content of the element in which cascade K→L transitions take place at $\lambda<\lambda_K$ the ionisation of L-shell by the K-radiation of other atoms of the same element (the process of self-excitation was investigated in section 8.6.2) is important. The contribution of this process is not so large as the contribution of cascade transitions and equals approximately 15–30%.

Experiments were carried out in [305] to confirm the effect of cascade transitions on the intensity of x-ray L-fluorescence. Radiation of an x-ray tube with a silver anode at a voltage of up to 50 kV excited the intensity of antimony ($Z = 51$) and lead ($Z = 82$) in a specimen with the composition 98% Sb_2O_5+2% PbO. The long-wave component of the primary bremsstrahlung of the tube was weakened by an aluminium filter with a thickness of 0.5 mm. The intensities of the x-ray L-fluorescence of antimony and lead were compared.

If it is approximately accepted that the intensity of fluorescence is proportional to $(V-V_L)^n$, where V_L is the ionisation potential of the L-shell, then the value of the ratio:

$$\frac{I_{\text{Sb}}^{L}}{I_{\text{Pb}}^{L}}=\left(\frac{V-V_{\text{L}}^{\text{Sb}}}{V-V_{\text{L}}^{\text{Pb}}}\right)^{n}=\left(1+\frac{V_{\text{L}}^{\text{Pb}}-V_{\text{L}}^{\text{Sb}}}{V-V_{\text{L}}^{\text{Pb}}}\right)^{n}$$

decreases with the increase of voltage V. This confirms the accurate calculations, carried out using equation (8.32) of the intensity determined by photoionisation of the L-shell. Selective excitation of the atoms of antimony by L-radiation of lead (this process was examined in section 8.6.3) for the given composition of the specimen was very low, according to the calculations.

Figure 8.8 compares the experimental and calculated dependences of the ratio of the intensities of the fluorescent radiation of antimony and lead on the x-ray tube voltage.

In Fig. 8.8 a voltage of 30.5 kV is characterised by a large change of the dependence of the relative intensity from decreasing to increasing. This voltage corresponds to the ionisation potential of the K-shell of antimony. The cascade transitions, formed at higher voltages, result in the transfer of vacancies to the L-shell and, correspondingly, to an increase of the intensity of the L-radiation of antimony. The energy of the primary photons is insufficient for ionisation of the K-shell of lead. The figure indicates that the calculations confirm the observed experimental dependence.

If the range of the wavelength between the K- and L-edges of adsorption includes the relatively high intensity part of the inhomogeneous primary radiation, resulting in effective ionisation of the L-shell, the role of the cascade processes in the formation of L-fluorescence rapidly decreases.

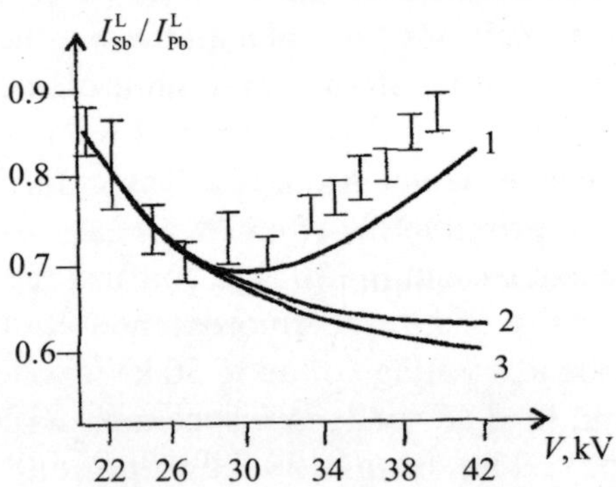

Fig. 8.8. The effect of cascade transitions on the intensity of Lα-radiation of antimony. 1) calculation of the ratio of the intensities of antimony and lead taking into account photoionisation, self-excitation and cascade transitions, 2) calculation of this ratio taking photoionisation and self-excitation into account, 3) calculation of this ratio taking only photoionisation into account. The line segments indicate the confidence ranges of experimental measurements.

8.5.2. Ionisation of atoms by photo- and Auger electrons

The interaction of x-ray radiation with matter is accompanied unavoidably by the formation of photo- and Auger electrons in matter. The energy of these electrons may prove to be sufficient for the ionisation of the atoms of the irradiated specimens intensifying x-ray fluorescence. The first theoretical evaluation of the ionising effect of photoelectrons [307], carried out for a copper specimen at bremsstrahlung of an x-ray tube and a voltage of 30–40 kV shows that the contribution of this process to the total intensity of fluorescence of copper is expressed in fractions of a percent. However, later calculations [308–310], carried out for the case in which the energy of primary radiation photons is considerably higher than the ionisation potential of the atoms of the fluorescing element and also taking into account the ionisation of the atoms by the Auger electrons, resulted in the value of the examined contribution which varied in different cases from several to tens of percent. A relatively simple justification of the value of the contribution of photo- and Auger electrons to the formation of x-ray fluorescence of elements with low atomic numbers in monochromatic primary radiation was described in [311].

To simplify theoretical calculations, it is assumed that the excitation of atoms by photo- and Auger electrons takes place in the immediate vicinity of the atoms generating these electrons. The assumption on the formation of a secondary photon at the point of absorption of the primary radiation photon is based on the considerably lower penetrating capacity of the electrons in comparison with the penetrating capacity of x-ray radiation.

The process of transformation of the primary photon to the photon of x-ray fluorescence through the ionising effect of photo- and Auger electrons will be examined.

Examination starts with photoelectrons.

The probability of ionisation of the atoms of some element j by primary radiation is determined by the equation:

$$\tau_{mj}(\lambda)c_j,$$

where $\tau_{mj}(\lambda)$ is the partial coefficient of true absorption of primary radiation by the m-level of the element j, c_j is the content of this element in the specimen.

Since every ionisation is accompanied by the appearance of a photoelectron, this probability is also the probability of formation of the photoelectron of the m-level.

Let $n_{mj}(c_i, E)$ be the probability of ionisation by the photoelectron with the energy E of the atoms of the element i whose content in the specimen is c_i. This probability for a monoelement specimen is determined by the expression (1.16), and for a multicomponent material by a more complicated dependence (see [20, 312]). Consequently, the probability of ionisation of the atoms of element i by the photoelectrons, formed under the effect of primary radiation, is equal to:

$$\tau_{mj}(\lambda)c_j n_{mj}(c_i, E).$$

Multiplying the resultant value by the probability of x-ray fluorescence yield ω_q and the probability p_{qi} of emission of some line of the q-series of element i, the following equation is obtained for the required probability $\vartheta_i(\lambda)$ of transformation of the primary photon to a fluorescent one:

$$\vartheta_i(\lambda)=\omega_q p_{qi}\tau_{mj}(\lambda)c_j n_{mj}(c_i, E). \tag{8.34}$$

Taking into account the fact that a fluorescing material always contains several elements j, ionised by primary radiation, and the photoelectrons are formed during ionisation of different m-shells of j-th atoms, equation (8.34) is presented in the following form:

$$\vartheta_i(\lambda)=\omega_q p_{qi}\sum_j c_j \sum_m \tau_{mj}(\lambda)n_{mj}(c_i, E), \tag{8.35}$$

where summation is carried out with respect to all elements j, present in the specimen, and with respect to all m-levels of these elements.

In summation, it should be taken into account that the process of ionisation of the atoms i by photoelectrons can take place at the energy of the primary photon sufficient not only for ionisation of the atoms of the element j but also for providing the energy higher than the binding energy of the electrons in the atoms of the element i to the produced photoelectrons. This means that the following condition must be satisfied:

$$E > E_m^j + E_q^i,$$

or in wavelengths

$$\lambda < \frac{\lambda_m^j \lambda_q^i}{\lambda_m^j + \lambda_q^i}.$$

The excitation of atoms with Auger electrons will now be examined.

The difference between the Auger electrons and photoelectrons is that:

1. Their energy is independent of the energy of primary radiation photons. Therefore, $E > E_j^m$, i.e., the energy of primary quanta should be sufficient for the ionisation of the atoms emitting Auger electrons.

2. The Auger electrons of some of the elements present in the material do not have sufficient energy sufficient for the ionisation of the atoms of the element whose fluorescence is investigated. Therefore, summation with respect to j in equation (8.35) relates only to the elements whose Auger electrons are capable of ionising the q-shell of the atoms i.

3. The number of the Auger electrons is smaller by a factor of $(1-\omega_j)$ than the number of the ionised atoms, where ω_i is the yield of fluorescence of element j.

4. The energy of the Auger electrons is determined by the difference of the energies of the initial and final levels of the transition. The energies of the sublevels inside the shell differ only slightly. Since they do not introduce any significant error, for the Auger K-electrons consideration can be restricted to the average K-LL- and K-LM-transitions, and only the transitions L-MM can be examined for the Auger L-electrons.

For example, in ionisation of the atoms by the Auger K-electrons, the introduction of the appropriate corrections into equation (8.35) results in the following dependence for the probability $\vartheta_i(\lambda)$ of the transformation of primary radiation photons to x-ray fluorescence photons:

$$\vartheta_i(\lambda) = \omega_q p_{qi} \sum_j c_j \tau_j(\lambda)(1-\omega_j)\left(0.7 n_j(c_i, E_{KLL}) + 0.3 n_j(c_i, E_{KLM})\right), \quad (8.36)$$

where the coefficients 0.7 and 0.3 characterise the relationship of the K-LL- and K-LM transitions.

Using cquation (8.35), the equations (8.36) and (8.32) may be used to calculate the intensity determined by the excitation of the atoms of the element i by the photo- and Auger electrons formed in the irradiated material.

When studying the process of excitation of x-ray radiation by an electron flux (primary spectra) it was found that the ionisation cross-section of the electrons rapidly increases with the increase of the energy and reaches a maximum when this energy is considerably higher than the ionisation potential. Therefore, it is efficient to excite x-ray radiation by photo- and Auger electrons with the energy considerably higher than the binding energy of the atomic electrons.

As an example, attention will be given to the excitation of sodium atoms ($E_K = 1.07$ keV) by monochromatic radiation of the ^{109}Cd radioactive source with the energy of 22.1 keV. It is assumed that the investigated specimens contain 0.1% of sodium and 99.9% of another element, referred to as the filler. Figure 8.9 shows the ratio of the intensity of x-ray fluorescence of sodium, excited by photoelectrons (I_i^P) and Auger electrons (I_i^A), to its intensity formed in direct ionisation of its atoms by primary radiation

$$\left(I_i^{\text{rel}} = \frac{I_i^P + I_i^A}{I_i} \right).$$

The results, presented in Fig. 8.9, show that in the given conditions, the intensity of x-ray fluorescence of sodium, excited by photo- and Auger electrons, can be 1.5–2 times higher than the intensity formed as a result of the ionisation of the atoms of sodium by primary radiation. The role of the photo- and Auger electrons in the excitation of fluorescence depends strongly on the atomic number Z_f of the element-filler. For a multicomponent specimen, this role is determined by the element composition of the fluorescing material.

The appearance of maxima in Fig. 8.9 is associated with the fact that the increase of the atomic number of the element-filler increases the probability of ionisation of its atoms by primary x-ray radiation in proportion to the coefficient of photoelectric adsorption. At the same time, the number of photoelectrons of this element increases. This is accompanied by a decrease of the energy of the photoelectrons and, consequently, the number of ionisations $n(E)$ per electron. The first factor is dominant in the case of the small atomic numbers, and the role of the photoelectrons becomes more important with the increase of Z_f. For high atomic numbers, the second factor is more important and the effect of the photoelectrons decreases with increasing Z_f, and tends to zero at the photoelectron energies close to the binding energy of the K-electrons in the Na atoms.

The maximum for the L-photoelectrons is greatly shifted in relation to the maximum of the K-photoelectrons to the area of high atomic numbers of the elements-fillers. The shift is explained by the low probability of photoelectric absorption of high energy x-ray photons by the L-shell. The high value of the contribution is explained by the higher 'population' of the L-shell by the electrons.

The completely different dependence on the atomic number of the element-filler is detected for the contribution of the Auger electrons. The energy of these electrons increases with increasing Z_f and the ionisation

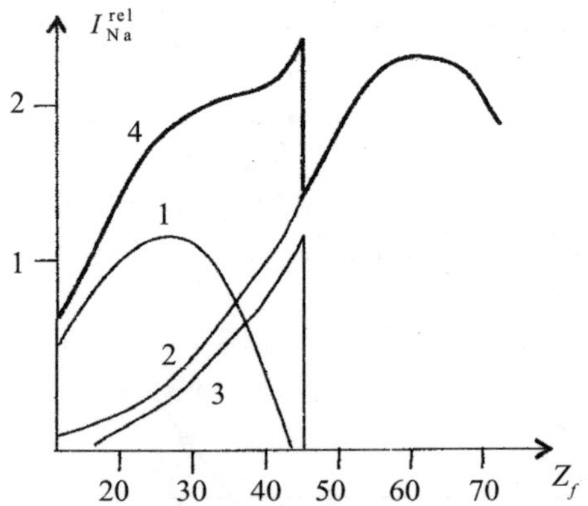

Fig. 8.9. The role of photo- and Auger electrons of the elements with the atomic number Z_f in the excitation of sodium atoms (C_{Na} = 0.1%) in monochromatic primary radiation with the photon energy of 22.1 keV. 1) the contribution of K-photoelectrons, 2) the contribution of L-photoelectrons, 3) contribution of Auger K-electrons, 4) the total contribution of the photo- and Auger electrons.

capacity of the electrons increases correspondingly. (In the case of small atomic numbers, the energy of the Auger electrons may prove to be insufficient for the ionisation of the K-shell of the Na atoms). An increase of Z_f is also accompanied by an increase of the number of ionisations and, consequently, of the number of the Auger electrons. Of course, the factor which reduces this number in proportion with the difference $1-\omega_i$ decreases at the same time. This factor is not so important, and the resultant value increases with the increase of Z_n to the maximum value at some atomic number and subsequently rapidly decreases to zero when the binding energy of the electrons of the K-level of the element-filler is higher than the energy of the primary photons.

It should be mentioned that the increase of the atomic number of the fluorescing element results in a rapid decrease of the contribution of photo- and Auger electrons to the ionisation of the K-shell. On the other hand, for the elements with small atomic numbers (beryllium, boron, carbon, oxygen, nitrogen) this contribution is controlling (see section 8.10).

8.6. Processes of excitation of x-ray fluorescence for the case in which the areas of interaction of primary radiation and formation of the fluorescent photon differ

The diagram of this process is shown in Fig. 8.10. The primary radiation photon, incident under the angle φ on the surface of the specimen, interacts at the point A in the depth x with the atom matter and transforms to an intermediate carrier of its energy (for example, the photon of scattered radiation or fluorescent radiation of some element). At some point B in the depth x_1 this energy carrier transforms to the photon of the fluorescent radiation (in which we are interested) emitted in the direction of the detector under the angle ψ to the surface of the specimen.

The intensity of x-ray fluorescence in the investigated case is determined by integration with respect to the depths x and x_1, and also to the directions

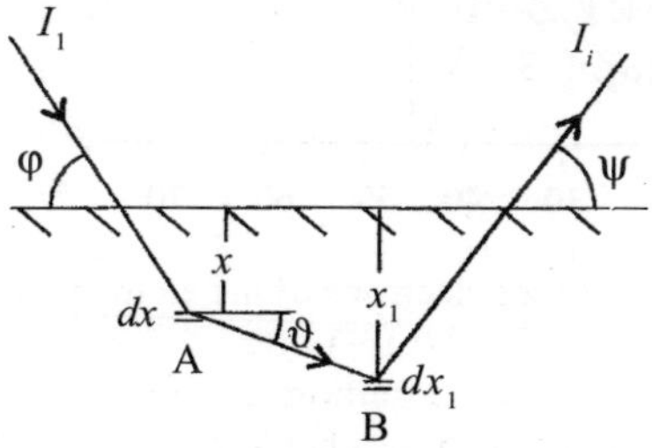

Fig. 8.10. Two-stage transformation of a primary photon to a fluorescent one.

of propagation of the intermediate energy carrier, given by the value of the angle ϑ.

For the massive specimen, assuming the isotropic nature of the propagation of intermediate radiation, the intensity of fluorescence is determined by the same equation (8.32) where the probability $\vartheta_i(\lambda)$ of transformation of the primary quantum to a fluorescent one has the form:

$$\vartheta_i(\lambda) = P_1 L P_2, \tag{8.37}$$

Here P_1 is the probability of transformation of the primary photon to an intermediate energy carrier of the photon, P_2 is the probability of transformation of this carrier to the fluorescent radiation photon, recorded by the detector. Value L is some transfer function, determined by the expression [291]:

$$L = \frac{1}{2}\left(\frac{\sin\varphi}{\mu(\lambda)}\ln\left(1+\frac{\mu(\lambda)}{m\sin\varphi}\right)+\frac{\sin\psi}{\mu(\lambda_i)}\ln\left(1+\frac{\mu(\lambda_i)}{m\sin\psi}\right)\right). \tag{8.38}$$

Function L depends on the value m which characterises the attenuation in the matter of the flux of intermediate energy carriers of primary radiation.

If it is assumed that $\mu(\lambda) \ll m\sin\varphi$ and $\mu(\lambda_i) \ll m\sin\psi$, then the values of the logarithms in the equation (8.38) may be transformed: $\ln(1-x) \approx x$, and function L is greatly simplified:

$$L \approx \frac{1}{m}. \tag{8.39}$$

Expression (8.39) is a rough approximation and can be used only in semi-quantitative calculations.

If the non-saturated layer of the specimens fluoresces, the equation for the probability $\vartheta_i(\lambda)$ is more complicated [313–315].

8.6.1. Effect of scattering processes on the intensity of x-ray fluorescence

These processes usually contribute to the total intensity of x-ray fluorescence by the value not exceeding units of a percent. However, in extreme conditions (efficiently scattering filler, short-wave primary and fluorescent radiation), the contribution may reach 10–20%. When modelling these processes, considerations are usually restricted to assuming that scattering is isotropic [316–318], although the previous (chapter 5) study of the scattering processes shows that the intensity of scattered radiation depends strongly on the scattering direction.

The possible variants of manifestation of the scattering effects in the formation of the intensity of fluorescent radiation will be examined separately.

a. X-ray fluorescence is excited coherently and non-coherently by scattered primary radiation.

For the case of coherent scattering of primary radiation at the point A (Fig. 8.10), the parameters of the expressions (8.37) and (8.38) have the following values:

$$P_1 = \sigma_{\mathrm{coh}}(\lambda), \;\; P_2 = \tau_{iq}(\lambda)\omega_q p_{iq} c_i, \;\; m = \mu(\lambda). \tag{8.40}$$

For the case of non-coherent scattering of primary radiation at the point A, the parameters of the equations (8.37) and (8.38) are as follows:

$$P_1 = \sigma_{\mathrm{nc}}(\lambda), \; P_2 = \tau_{iq}(\lambda_{\mathrm{nc}})\omega_q p_{iq} c_i, \;\; m = \mu(\lambda_{\mathrm{nc}}). \tag{8.41}$$

b. The x-ray fluorescence, formed in the specimen, is scattered coherently and non-coherently in the given direction.

In the case of coherent scattering of fluorescent radiation at point B, the parameters of the relationships (8.37) and (8.38) have the following values:

$$P_1 - \tau_{iq}(\lambda)\omega_q p_{lq} c_i, \; P_2 = \sigma_{\mathrm{ooh}}(\lambda_i), \;\; m = \mu(\lambda_i). \tag{8.42}$$

If the resolution capacity of the spectrometer is not high and relatively long-wave fluorescent radiation is recorded, in addition to the coherently scattered radiation the non-coherently scattered fluorescent radiation is also recorded for which:

$$P_1 = \tau_{iq}(\lambda)\omega_q p_{iq} c_i, \;\; P_2 = \sigma_{\mathrm{nc}}(\lambda_i), \;\; m = \mu(\lambda_i). \tag{8.43}$$

In this case, the non-coherently scattered fluorescent radiation is incident on the main line, distorting its shape on the long-wave side.

The short-wave fluorescent radiation, non-coherently scattered by the specimen, represents a separate broadened maximum situated on the long-wave side in relation to the peak of coherently scattered radiation (Fig. 5.13) and is not recorded by measuring equipment.

More accurate equations for taking into account the effect of the scattering processes on the intensity of x-ray fluorescence also take into account the anisotropy of scattered radiation [319]. In [320], in examining the scattering processes, the authors used the Boltzmann transport equation. This resulted in a better agreement with the experimental data. Since the contribution of the scattering effects to the excitation of fluorescence is usually not large, the equations presented in the sections are sufficient for taking this contribution into account.

8.6.2. Ionisation of the L-shell by K-radiation of the same element (self-excitation)

If the energy of the primary ions is sufficient for the ionisation of the K-level of some element, then the K-radiation of this element is capable of exciting its L-shell. The intensity of x-ray fluorescence, formed in this process, is determined by the equation (8.32), in which the probability of transformation ϑ_i of primary radiation to fluorescent L-radiation is given by the equations (8.37) and (8.38) where $P_1 = \tau_{iK}(\lambda)\omega_K c_i$ (it is assumed that all the lines of the K-series take part in the ionisation of the L-shell. Therefore $\sum_j p_{iK} = 1.$):

$$P_2 = \tau_{iL}(\lambda_K)\omega_L p_{iL} c_i,$$
$$m = \mu(\lambda_K).$$

The role of self-excitation in the formation of L-fluorescence was investigated in [321]. In section 8.5.1, it was noted that in monochromatic primary radiation with the wavelength smaller than the K-edge of adsorption, the contribution of the self-excitation process to the ionisation of the L-shell may be considerable.

8.6.3. Excitation of x-ray fluorescence by radiation of other elements, present in the irradiated material

In a multicomponent specimen, this process is almost completely unavoidable and is referred to as the selective excitation effect. The approximate equation for the selectively excited intensity of x-ray fluorescence in monochromatic primary radiation was derived in 1952 (see [322]) and was made more accurate by more rigorous derivation in 1954 (see [291, 323]).

In selective excitation, the parameters of the equations (8.37) and (8.38) have the following values:

$$P_1 = \tau_{qj}(\lambda)\omega_{qj} c_j, \quad P_2 = \tau_{qi}(\lambda_j)\omega_{qj} p_{qik} c_i, \qquad m = \mu(\lambda_j), \qquad (8.44)$$

where P_1 is the probability of formation of a primary photon in the sum of the photons of all lines of the q-shell of element j, i.e., it is assumed that all the lines of the element j take part in the excitation of the q-shell of the atoms i. This assumption is made to simplify the following explanation. In accurate calculations, the excitation of the atoms of the element i by the photons of every line of the element j is examined separately, followed by summation of the results.

The contribution of the investigated process to the intensity of x-ray fluorescence is determined by the ratio of the intensity I_i^D, excited by the radiation of the element j, to the intensity I_i, excited directly by the primary photons:

$$\frac{I_i^D}{I_i}=\frac{\vartheta_i^D(\lambda)}{\vartheta_i(\lambda)}=c_j K_{ij}\omega_{qj}L, \tag{8.45}$$

where $K_{ij}=\dfrac{\tau_{qj}(\lambda)\tau_{qi}(\lambda_j)}{\tau_{qi}(\lambda)}$.

The investigated ratio may be evaluated approximately with respect to using the equation (8.39) for the value L which in the investigated case (see [268]) is determined as: $L\approx\dfrac{\xi}{\mu(\lambda_j)}$, where ξ changes by a factor of 1.5 and more [324] with the variation of the composition of the specimen, the content of the exciting element and the primary radiation wavelength. In this case:

$$\frac{I_i^D}{I_i}\approx K_{ij}\omega_{qj}\frac{\xi c_j}{\mu(\lambda_j)}. \tag{8.46}$$

Equation (8.46) shows that the contribution of the effect of selective excitation to the intensity of x-ray fluorescence increases with the increase of the content of the exciting element and the decrease of the coefficient of attenuation of radiation of this element in the specimen.

The magnitude of the maximum contribution of the investigated process is estimated for the low content of iron in a nickel specimen ($c_{Fe}\to 0$, $c_{Ni}\to 1$, $\lambda_{Fe}=1.94$ Å, $\lambda_{Ni}=1.66$ Å, $\omega_{Ni}=0.41$). The wavelength of primary radiation is assumed to be equal to the wavelength of the Zn Kα-line. The value of L was equal to $L=0.0069$. Substituting these numerical values into equation (8.45) gives for the investigated case:

$$\frac{I_i^D}{I_i}\approx 1.05.$$

The resultant value indicates that the intensity of x-ray fluorescence of iron, formed as a result of excitation of iron atoms by nickel radiation, is close to the fluorescence intensity of iron, directly excited by primary radiation.

The effect of selective excitation is so strong that it is almost always taken into account in the calculations of the intensity of x-ray fluorescence. This means that it is assumed that:

$$(I)_i=I_i\left(1+\frac{I_i^D}{I_i}\right)=I_i(1+\gamma_i), \tag{8.47}$$

where the value of γ_i is determined by equation (8.45).

For several selectively exciting elements, the value γ_i is determined by the summation of the examined effects for these elements.

8.6.4. Third order effects

These effects assume the transformation of a primary photon to a fluorescent one in three consecutive interactions. The most significant interaction is the photoelectric one, in which primary radiation excites atoms of some element C, the radiation of element C excites the atoms of element B whose radiation in turn excites the atoms of the relevant element A (Fig. 8.11).

The expression for the intensity of x-ray fluorescence, formed as a result of the third order effects, is derived by integrating with respect to the depth of distribution x_I, x_{II}, x_{III} of the atoms of the elements C, B, A, and also with respect to the direction of propagation of radiation of the atoms C and B.

The intensity of x-ray fluorescence, determined by the third order photoelectric effect, is defined by the same equation (8.32) in which the probability $\vartheta_i(\lambda)$ of transformation of a primary radiation photon to a fluorescent radiation photon of element A is determined by the equation:

$$\vartheta_i(\lambda) = P_1 P_2 P_3 M, \tag{8.48}$$

where M is some transfer function determined by a relatively complicated equation, including several integral logarithms. The approximate value of M may be presented (see [324]) in the form:

$$M \approx \frac{1}{4}\left(\frac{\sin\varphi}{\mu(\lambda)}\ln\left(1+\frac{\mu(\lambda)}{m\sin\varphi}\right)+\frac{\sin\psi}{\mu(\lambda_i)}\ln\left(1+\frac{\mu(\lambda_i)}{m\sin\psi}\right)\right)^2. \tag{8.49}$$

As shown in [325], the approximation error of (8.49) is minimum at $m = 0.7\mu(\lambda_B) + 0.3\mu(\lambda_C)$.

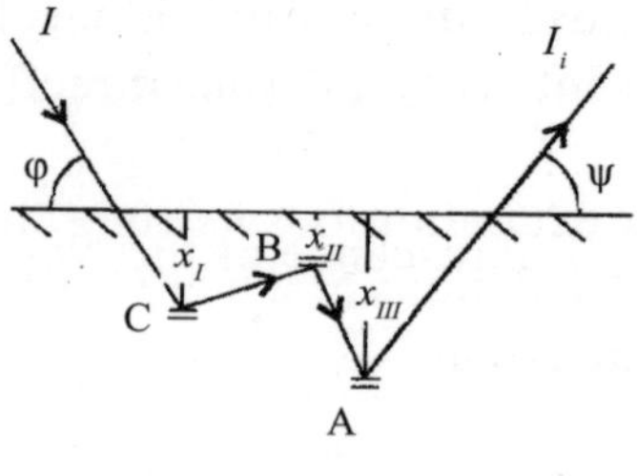

Fig. 8.11. The diagram of formation of x-ray fluorescence as a result of the third order effects.

The second variant was proposed in [326]:

$$M \approx \frac{1}{4}\left(\frac{\sin\varphi}{\mu(\lambda)}\ln\left(1+\frac{\mu(\lambda)}{\mu(\lambda_B)\sin\varphi}\right)+\frac{\sin\psi}{\mu(\lambda_i)}\ln\left(1+\frac{\mu(\lambda_i)}{\mu(\lambda_B)\sin\psi}\right)\right)\times$$

$$\times\left(\frac{\sin\varphi}{\mu(\lambda)}\ln\left(1+\frac{\mu(\lambda)}{\mu(\lambda_C)\sin\varphi}\right)+\frac{\sin\psi}{\mu(\lambda_i)}\ln\left(1+\frac{\mu(\lambda_i)}{\mu(\lambda_C)\sin\psi}\right)\right). \tag{8.50}$$

The probabilities P_1, P_2 and P_3 of successive transformation of the photons have the following values in equation (8.48):

$$P_1 = \tau_{q\mathrm{C}}(\lambda)\omega_{qC} p_{q\mathrm{C}i} c_{\mathrm{C}},$$

$$P_2 = \tau_{q\mathrm{B}}(\lambda_{\mathrm{C}})\omega_{q\mathrm{B}} p_{q\mathrm{B}i} c_{\mathrm{B}},$$

$$P_3 = \tau_{q\mathrm{A}}(\lambda_{\mathrm{B}})\omega_{q\mathrm{A}} p_{q\mathrm{A}i} c_{\mathrm{A}}.$$

For several combinations of the elements, forming the composition of the specimen and causing the third order photoelectric effects, the total contribution of these elements is determined by the summation of these effects.

The maximum contribution of the photoelectric effect of the third order to the intensity of x-ray fluorescence is 3–5%. Therefore, it is completely acceptable to use the relationships (8.49) and (8.50) to take this effect into account for the value M.

The contribution of other third order effects (scattering–scattering-excitation of fluorescence, scattering–excitation–excitation, etc.) equals usually fractions of percent and is not considered in calculations in practice. Accurate calculations of the third order effects, taking the scattering processes into account, are complicated because of the anisotropy of scattered radiation.

8.7. Selective absorption of primary radiation

Let the composition of the fluorescent specimen to contain some element B whose absorption edge is situated between the wavelength λ of primary radiation and the wavelength λ_i of fluorescent radiation of element A (Fig. 8.12).

The intensity of x-ray fluorescence of the element A is determined by the same equation (8.32):

$$(I_i^{\mathrm{A}})' = \frac{k\vartheta_i(\lambda)}{\mu(\lambda)+m\mu(\lambda_i)}, \tag{8.51}$$

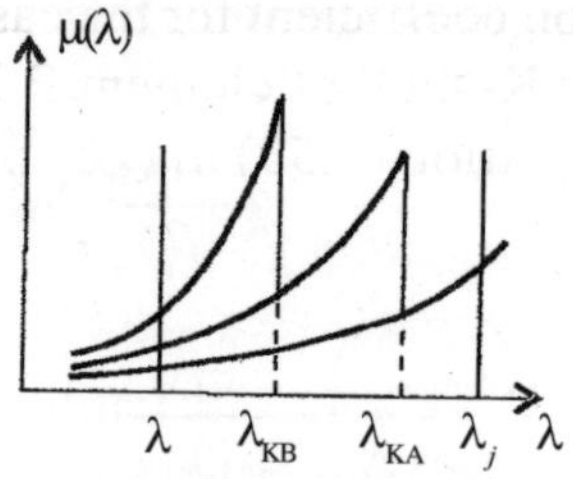

Fig. 8.12. Selective absorption of primary radiation with wavelength λ by element B.

where $m = \dfrac{\sin\varphi}{\sin\psi}$.

The mass coefficient of attenuation $\mu(\lambda)$ in equation (8.51) takes into account the selective absorption of primary radiation by the atoms of element B. In fact, the investigated attenuation coefficient may be expressed by the attenuation coefficient of fluorescent radiation of element A by the dependence:

$$\mu(\lambda) = \left(\frac{\lambda}{\lambda_i}\right)^n \left(c_A\mu_A(\lambda_i)(S_K^A - 1) + c_B\mu_B(\lambda_i)(S_K^B - 1) + \mu(\lambda_i)\right), \quad (8.52)$$

where in comparison with equation (8.23), which is valued in the absence of element B, there is another term $c_B\mu_B(\lambda_i)(S_K^B - 1)$, determined by selective absorption.

The effect of the presence of element B on the intensity of x-ray fluorescence of element A will be evaluated. The measure of this effect is the relationship:

$$\frac{\Delta I_i}{\left(I_i^A\right)^N} = \frac{\left(I_i^A\right)^N - \left(I_i^A\right)'}{\left(I_i^A\right)^N},$$

where $(I_i^A)^N$ is the intensity of x-ray fluorescence of element A in the absence of selective absorption of primary radiation by element B, but with the constant attenuation coefficient $\mu(\lambda_i)$ of fluorescent radiation of element A.

Substituting the intensity values from (8.51) and (8.32) into the resultant equation:

$$\frac{\Delta I_i}{\left(I_i^A\right)^N} = \frac{\mu(\lambda) - \mu'(\lambda)}{\mu(\lambda) + m\mu(\lambda_i)} \quad (8.53)$$

where $\mu'(\lambda)$ is the attenuation coefficient for the case in which the absorption of primary radiation by the K-shell of element B is ignored.

The numerator of the equation (8.53) may be presented in the following form:

$$\mu(\lambda)-\mu'(\lambda)=c_{\mathrm{B}}\tau_{\mathrm{KB}}(\lambda),$$

where $\tau_{\mathrm{KB}}(\lambda)$ is the partial coefficient of photoelectric absorption of the primary radiation by the K-shell of element B. (Scattering of radiation in the K-shell is not taken into account).

Consequently, equation (8.53) gives the following equation for the attenuation of the intensity of x-ray fluorescence as a result of the selective absorption of primary radiation by element B:

$$\frac{\Delta I_i}{(I_i^{\mathrm{A}})^N}=\frac{c_{\mathrm{B}}\tau_{q\mathrm{B}}(\lambda)}{\mu(\lambda)+m\mu(\lambda_i)}, \tag{8.54}$$

i.e. the magnitude of the effect increases with the increase of the content of the element which causes selective absorption.

The previously investigated case of a low iron content will again be considered ($c_{\mathrm{Fe}}\to 0$, $\lambda_{\mathrm{Fe\ K}\alpha}=1.937$ Å in nickel ($c_{\mathrm{Ni}}\to 1$). In this case $\mu(\lambda)\approx\mu_{Ni}(\lambda)$ and $\mu(\lambda_{\mathrm{Fe}})\approx\mu_{\mathrm{Ni}}(\lambda_{\mathrm{Fe}})$. Let the wavelength of primary radiation be equal to the wavelength of the Kα-line of zinc (λ = 1.435 Å). It is also assumed that $m=\sin\varphi/\sin\psi=1$.

Taking into account that the partial and total absorption coefficients are linked by the relationship $\tau_{q\mathrm{B}}=\frac{S_{qB}-1}{S_{qB}}\tau_{\mathrm{B}}$, gives:

$$\frac{\Delta I_i^{\mathrm{Fe}}}{\left(I_i^{\mathrm{Fe}}\right)^N}=\frac{\tau_{\mathrm{Ni}}\left(1.435\right)\frac{7.98-1}{7.98}}{\mu_{\mathrm{Ni}}\left(1.435\right)+\mu_{\mathrm{Ni}}\left(1.937\right)}=0.73.$$

Thus, the effect of selective absorption on the intensity of x-ray fluorescence may prove to be very strong and comparable with the selective excitation of x-ray fluorescence.

8.8. Perturbing effect of elements on the intensity of x-ray fluorescence

The selective excitation effect is always accompanied by the selective absorption of primary x-ray radiation. Selective excitation increases the intensity of x-ray fluorescence of the element, and the selective absorption of primary radiation lowers this intensity. In fact, the exciting element transforms [327] the primary radiation energy adsorbed by it to its characteristic radiation exciting the atoms of the relevant element, with the transformation coefficient lower than 1 (the fluorescence yield ω_B is always lower than unity). At the same time, in this transformation the short-wave primary radiation is replaced by fluorescent radiation with a greater wavelength which is more effective in exciting the characteristic fluorescent radiation of the element under consideration. The total effect may prove to be both higher and lower than unity. This total effect will be referred to as the perturbation, and the elements causing this effect the perturbing elements. The wavelengths of the absorption edges of the perturbing elements are shorter than the wavelength of the characteristic line of the investigated element.

The intensity of x-ray fluorescence I_i^A of element A in the presence of the perturbing element B is determined by the equation:

$$I_i^A = (I_i^A)' + I_i^D, \tag{8.55}$$

where the intensity $(I_i^A)'$ is given by the equation (8.51), and I_i^D is the intensity selectively excited by the fluorescent radiation of element B (section 8.6.3).

Equation (8.55) may be transformed to the form:

$$I_i^A = (I_i^A)^N + (I_i^A)' - (I_i^A)^N + I_i^D = (I_i^A)^N \left(1 + \frac{(I_i^A)' - (I_i^A)^N}{(I_i^A)^N} + \frac{I_i^D}{(I_i^A)^N}\right). \tag{8.56}$$

The resultant dependence (8.56) links the total I_i^A and non-perturbed $(I_i^A)^N$ intensities.

In the investigated dependence, the ratio $\frac{\Delta I_i}{(I_i^A)^N} = \frac{(I_i^A)' - (I_i^A)^N}{(I_i^A)^N}$, given by the equation (8.54), is negative; the ratio $\frac{I_i^D}{(I_i^A)^N}$ may be easily obtained from the equation (8.45) by multiplying the ratio by $\frac{I_i}{(I_i^A)^N} = \frac{\mu'(\lambda) + m\mu(\lambda_i)}{\mu(\lambda) + m\mu(\lambda_i)}$:

$$\frac{I_i^D}{(I_i^A)^N} = c_B K_{BA} \omega_{qB} L(\lambda) \frac{\mu'(\lambda) + m\mu(\lambda_i)}{\mu(\lambda) + m\mu(\lambda_i)}. \tag{8.57}$$

Taking the equations (8.54) and (8.57) into account, the dependence (8.56) may be presented in the form:

$$I_i^A = (I_i^A)^N (1 + c_B \beta_B). \tag{8.58}$$

Coefficient β_B determines the value of the correction for the perturbing effect of element B introduced by the element into the non-perturbed intensity of element A. Its value is determined by the equation:

$$\beta_B = \frac{\tau_{qB}(\lambda)}{\mu(\lambda) + m\mu(\lambda_i)} \left(\frac{\tau_{qA}(\lambda_B)}{\tau_{qA}(\lambda)} \omega_{qB} L(\lambda)\left(\mu'(\lambda) + m\mu(\lambda_i)\right) - 1 \right). \quad (8.59)$$

This equation shows that the correction β_B depends strongly on the wavelength of primary radiation, the elemental composition of the investigated material and on the position of the specimen, i.e., on the angle of incidence of primary radiation and selection of fluorescent radiation. Therefore, by selecting the conditions of perturbation of x-ray fluorescence (the wavelength of primary radiation and the position of the specimen), it is possible to influence strongly the perturbation effect, up to its complete disappearance [327, 328].

The wavelength of primary radiation at which the presence of the perturbing element B in the specimen has no effect on the intensity of x-ray fluorescence of element A will be determined. In this case, the correction for the perturbing effect β_B should be equal to 0. Consequently, from (8.59) we obtain:

$$\frac{\tau_{qA}(\lambda_B)}{\tau_{qA}(\lambda)} \omega_{qB} p_{qBi} L(\lambda)\left(\mu'(\lambda) + m\mu(\lambda_i)\right) = 1.$$

From this, taking into account that $\frac{\tau_{qA}(\lambda_B)}{\tau_{qA}(\lambda)} \approx \left(\frac{\lambda_B}{\lambda}\right)^3$ and $L(\lambda) \approx \frac{\varepsilon}{\mu(\lambda_B)}$, and after elementary transformations:

$$\lambda \approx \lambda_i \sqrt[3]{\frac{m\omega_{qB}\varepsilon}{1-\omega_{qB}\varepsilon} \frac{1}{W_A(S_A - 1) + 1}},$$

where $W_A = \frac{C_A \mu_A(\lambda_A)}{\mu(\lambda_A)}$ is the factor of chemical composition (the equation (8.24) in section 8.3).

This relationship shows that the wavelength of primary radiation, resulting in the mutual compensation of the processes of additional excitation and selective excitation of exciting radiation, depends on the atomic numbers of the elements A and B, on the composition of the specimen and on the geometrical factor *m*.

The previously investigated example of the low iron content ($c_{Fe} \to 0$) in nickel ($c_{Ni} \to 1$) will again be examined. The fluorescence yield $\omega_{Ni} = 0.414$, the geometrical factor *m* is assumed to be equal to 1. If $c_{Fe} \to 0$, then also $W_{Fe} \to 0$. Coefficient ε changes in the range from 1.1 to 0.6 (decreases with an increase of the content of element B and with a decrease of the content of element A). For a low iron content in nickel ε = 0.6 [324].

Using these values in the calculations gives the wavelength of primary radiation resulting in the compensation conditions and equal to λ = 1.30 Å.

This means that the determined wavelength is a real value which is smaller than the wavelength of the absorption edge of nickel ($\lambda_{\mathrm{K\,Ni}} = 1.48$ Å).

Taking into account the possibility that the specimen may contain several perturbing elements, and using equation (8.26) for the non-perturbed intensity of x-ray fluorescence, equation (8.58) may be presented in the following form:

$$I_i^{\mathrm{A}} = k_i \frac{W_i}{1+\mathrm{A}_i W_i}\left(1+\sum_j \beta_{\mathrm{A}j} C_j\right), \tag{8.60}$$

where summation is carried out with respect to the perturbing elements only.

If the selective excitation is determined by the relatively weak Kβ-line of element B, or does not form at all (the absorption edge of element B this position between the adsorption edge and the line of the elements A), the perturbing effect is determined by the process of selecting adsorption, and the correction β_{B} for the perturbing effect is negative.

The proposed model of formation of the intensity of x-ray fluorescence with the perturbing effect of the elements taken into account has been used in x-ray fluorescence analysis of multicomponent materials [329, 330].

8.9. Excitation of x-ray fluorescence by polychromatic primary radiation

X-ray tubes whose radiation is inhomogeneous with respect to the spectral composition are used widely for exciting x-ray fluorescence. This radiation contains characteristic and bremsstrahlung radiation (section 3.1). The ratio of these components depends on the atomic number of the anode material and on the tube voltage.

In the absence of selective effects, the equation for the intensity of x-ray fluorescence, formed as a result of photoelectric ionisation of the atoms, in inhomogeneous primary radiation is derived (see [331]) by integrating the equation (8.32) with respect to the wavelength from the shortwave boundary λ_0 of the bremsstrahlung spectrum to the wavelength λ_q of the q-absorption edge of the fluorescing specimen:

$$I_i = \frac{s\sin\varphi}{4\pi R^2} c_i \frac{S_q - 1}{S_q} \omega_{qi} p_{qij} \frac{1}{\lambda_i} \int_{\lambda_0}^{\lambda_q} \frac{\lambda I(\lambda)\tau_i(\lambda) d\lambda}{\mu(\lambda)+m\mu(\lambda_i)}, \tag{8.61}$$

The value $I(\lambda)$ in (8.61) is the spectral distribution of primary radiation, including the bremsstrahlung and characteristic components.

This equation is used widely in the calculations of the intensity of x-ray fluorescence in practice, in particular, in procedural investigations and in realisation of a number of methods of x-ray fluorescence analysis.

In the two following sections attention will be given to two possible variants of calculating the intensity of x-ray fluorescence, excited by inhomogeneous primary radiation, without carrying out integration with respect to the wavelength of this radiation.

8.9.1. The integral-free model of calculating the intensity of x-ray radiation

In the absence of absorption jumps of some elements (with the exception of the element whose fluorescence is investigated) in the range between the wavelength of monochromatic primary radiation and the wavelength of the fluorescence line for the intensity of x-ray fluorescence it is justified to use equation (8.26) in which the effect of elemental composition on the intensity is expressed through the factor of chemical composition W_i.

If primary radiation is inhomogeneous, after integration with respect to the wavelengths, equation (8.31) can be presented in the form (see [330]) adequate to the dependence (8.61):

$$I_i = K_i' W_i \int_{\lambda_0}^{\lambda_q} \frac{\lambda I(\lambda) \tau_i(\lambda) d\lambda}{(A_i W_i + 1)\left(m + \left(\frac{\lambda}{\lambda_i}\right)^n\right)}, \tag{8.62}$$

where $A_i = \dfrac{S_q - 1}{1 + m\left(\dfrac{\lambda_i}{\lambda}\right)^n}$ and $K_i' = \dfrac{s \sin\varphi}{4\pi R^2} \dfrac{1}{\lambda_i} \dfrac{S_q - 1}{S_q} \dfrac{\omega_q p_q}{\mu_i(\lambda_i)}$.

Attention will be given to some effective wavelength λ_{eff} of primary radiation and, using the generalised theorem of the mean, the value $(A_i W_i + 1)$ is placed behind the integral sign. From (8.62):

$$I_i = K_i' \frac{W_i}{A_i W_i + 1} \int_{\lambda_0}^{\lambda_q} \frac{\lambda I(\lambda) \tau_i(\lambda) d\lambda}{m + \left(\frac{\lambda}{\lambda_i}\right)^n}, \tag{8.63}$$

where $A_i = \dfrac{S_q - 1}{1 + m\left(\frac{\lambda_i}{\lambda_{eff}}\right)^n}$ and the wavelength λ_{eff} depends on the element composition of the fluorescing object.

If it is assumed that λ_{eff} = const, then A_i is constant (for the given conditions of measuring the fluorescence of the element) and equation (8.63) yields an equation identical with the previously examined equation (8.26) for monochromatic primary radiation:

$$I_i = K_i \frac{W_i}{1 + A_i W_i}, \tag{8.64}$$

where the coefficient $K_i = K_i' \int_{\lambda_0}^{\lambda_q} \frac{\lambda I(\lambda) \tau_i(\lambda) d\lambda}{m + \left(\frac{\lambda}{\lambda_i}\right)^n}$ is independent of the wavelength and element composition of the specimen.

Since λ_{eff} is a function of the element composition of the specimen, the parameter A also depends on this composition. It is assumed that A_i is a linear function of factor W, i.e. $A_i = a_i + b_i W_i$. Consequently:

$$I_i = K_i \frac{W_i}{1 + a_i W_i + b_i W_i^2}. \tag{8.65}$$

The values of the parameters a_i and b_i in (8.65) (or A_i in (8.64)) can be determined by the method of least squares, if the intensity I_i is calculated from the general equation (8.62) for a large set of factors W_i of the composition.

The error of calculating the intensity of x-ray fluorescence using the derived dependences (8.64) and (8.65) in relation to the calculations carried out using the general equation (8.62) are presented in Table 8.1. Here it is assumed that the inhomogeneous primary radiation is generated by an x-ray tube with a palladium anode with a beryllium window 0.3 mm thick. The tube voltage was 40 kV, the geometrical conditions: $\varphi = 50°$, $\psi = 35°$. The errors were estimated for the elements C, Mn, Fe, Ni, W for the following variations of the content of these elements: iron from 40 to 99%, chromium and nickel from 0.05 to 40%, manganese, molybdenum and tungsten from 0.01 to 20%. These variations correspond to the maximum possible variations in high-alloy steels.

Table 8.1 shows that the error of calculating the intensity of x-ray fluorescence decreases by almost an order of magnitude when passing from equation (8.64) to equation (8.65) and equals a value of the order of thousandths of a percent (for iron and chromium hundredths of a percent) which is considerably lower than the experimental error of measuring these intensities.

8.9.2. Equivalent parameters of inhomogeneous primary radiation

In a number of cases, to simplify the calculations of the intensity of x-ray

Table 8.1. Comparison of the mean relative errors of calculating the intensity of x-ray fluorescence using the integral-free approximate equations (8.64) at (8.65). Accurate calculations were carried out using equation (8.62)

Element	(8.64)		(8.65)		
	A_i	$(\Delta I_i/I_i)$, %	a_i	b_i	$(\Delta I_i/I_i)$, %
Cr	1.291	0.206	1.410	−0.358	0.0168
Mn	1.378	0.076	1.453	−0.444	0.0036
Fe	1.130	0.542	1.309	−0.170	0.0751
Ni	1.391	0.022	1.430	−0.446	0.0006
W	0.520	0.004	0.525	−0.044	0.0001

fluorescence, it is necessary to introduce some effective wavelength of primary radiation (determined from the theorem of the mean) which replaces the polychromatic primary radiation (see [26, 332]).

Another approach is based on searching for a monochromatic model which would minimise the dependence of the selected wavelength on the elemental composition of the specimens [333].

Generally speaking, there is an infinite number of pairs (monochromatic wavelength and its intensity) whose effect is equivalent to the effect of inhomogeneous primary radiation, i.e., they ensure the same intensity of x-ray fluorescence. From this large number a pair that is relevant to these investigations is selected by applying an additional condition: the error of calculating the intensity of x-ray fluorescence should be minimum at a large variation of the composition of the fluorescing material [334, 335].

Equation (8.62) for the intensity of x-ray fluorescence in inhomogeneous primary radiation may be presented in the form:

$$I_i = C_i W_i \int_{\lambda_0}^{\lambda_q} \frac{\lambda^4 I(\lambda) d\lambda}{\left(\frac{\lambda}{\lambda_i}\right)^3 (W_i(S_q - 1) + 1) + m}. \tag{8.66}$$

To simplify considerations, in (8.66) it is assumed that the coefficients of photoelectric absorption and attenuation of x-ray radiation are proportional to the cube of wavelength.

As a result of replacing the spectral distribution of inhomogeneous primary radiation by the values of the wavelength λ_1, equivalent to the inhomogeneous primary radiation, and by the intensity I_1 of the primary radiation of this wavelength, the intensity of x-ray fluorescence should remain constant. This means that:

$$I_i = C_i W_i \int_{\lambda_0}^{\lambda_q} \frac{\lambda^4 I(\lambda) d\lambda}{\left(\frac{\lambda}{\lambda_i}\right)^3 (W_i(S_q - 1) + 1) + m} =$$

$$= C_i W_i \frac{\lambda_1^4 I_1}{\left(\frac{\lambda_1}{\lambda_i}\right)^3 (W_i(S_q - 1) + 1) + m}. \quad (8.67)$$

For the two specimens whose composition is characterised by the factors W_{i1} and W_{i2}, the error of the monochromatic model does not occur, if the quantities $\lambda_1 = \lambda_\vartheta$ and $I_1 = I_\vartheta$ are the solution of the system of two equations of the type (8.67), written for these factors. The solution of this equation gives:

$$\lambda_\vartheta = \lambda_i \sqrt[3]{\frac{m\left(1 - \frac{t_1}{t_2}\right)}{\frac{t_1}{t_2}(W_{i1}(S_q - 1) + 1) - (W_{i2}(S_q - 1) + 1)}} \quad (8.68)$$

and

$$I_\vartheta = \frac{\left(\frac{\lambda_\vartheta}{\lambda_i}\right)(W_{i1}(S_q - 1) + 1) + m}{\lambda_\vartheta^4} t_1,$$

where $t_n = \int_{\lambda_0}^{\lambda_q} \frac{\lambda^4 I(\lambda) d\lambda}{\left(\frac{\lambda}{\lambda_i}\right)^3 (W_{in}(S_q - 1) + 1) + m}$, $(n = 1, 2)$.

Calculating the equivalent parameters λ_ϑ and I_ϑ for the set of doubled probes gives the mean values $\overline{\lambda_\vartheta}$ and $\overline{I_\vartheta}$.

If the values of the intensities, related to the intensity I_i^{st} of some standard are examined, then, as indicated by (8.67), these values are a function of only the equivalent wavelength:

$$R = \frac{I_i}{I_i^{st}} = \frac{W_i}{W_i^{st}} \frac{\left(\frac{\overline{\lambda}_\vartheta}{\lambda_i}\right)^3 (W_i^{st}(S_q - 1) + 1) + m}{\left(\frac{\overline{\lambda}_\vartheta}{\lambda_i}\right)^3 (W_i(S_q - 1) + 1) + m}. \quad (8.69)$$

The dependence of the error of calculations, using equation (8.69), on the relative intensity of x-ray fluorescence of the wavelength of the monochromatic model is presented in Fig. 8.13.

The monochromatic model was used for the case of inhomogeneous primary radiation of an x-ray tube with a palladium anode working at a

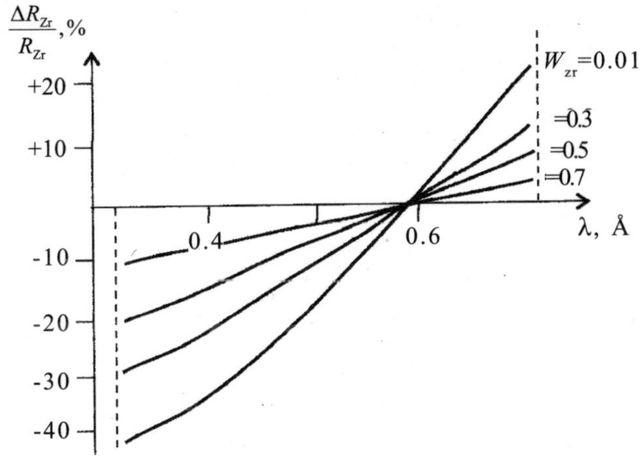

Fig. 8.13. Dependence of the error of calculating the relative intensity of Zr Kα-fluorescence radiation on the wavelength of the selected monochromatic approximation.

voltage of 40 kV. The angle of incidence of the primary radiation $\varphi = 48°$, the selection angle of x-ray fluorescence $\psi = 35°$. Zirconium was the fluorescing element. The factor of chemical composition changed in the range from 0.01 to 1. The dotted line indicates the short-wave boundary of bremsstrahlung and the wavelength of the absorption edge of zirconium.

Figure 8.13 shows that the error of calculating the relative intensities depends greatly on the selection of the wavelength which replaces the inhomogeneous primary spectrum. The minimum error (of the order of tenths of a percent) is detected at the equivalent wavelength (for zirconium in the given conditions $\overline{\lambda}_э = 0.569$ Å). For the effective wavelength determined from the mean value theorem ($\lambda_{eff} = 0.495$ Å) the error may reach 15%.

The investigated solution (in contrast to the theorem of the mean) does not impose any restrictions on the spectral composition of primary radiation and may be used if the composition of the radiation contains the characteristic component. Any series of the characteristic spectrum of the anode may be replaced by equivalent characteristics. For example, the introduction of the equivalent characteristics for K-radiation of copper, molybdenum and palladium anodes results in an error of the order of 0.05%. The replacement of eight lines of the L-series of the tungsten anode by equivalent characteristics results in the error of calculating the intensity of fluorescence of elements not exceeding 0.35% (see [336]).

8.9.3. Second order effects in inhomogeneous primary radiation

The complex spectral composition of the radiation of x-ray tubes results in large corrections in the value of the second order effect in the formation of x-ray fluorescence. Some of these effects will be investigated.

8.9.3.1. Cascade transitions and self-excitation

These second order processes, as mentioned previously, may influence the intensity of x-ray fluorescence of the lines of L-, M- and other series. The expression for the intensity, determined by cascade transitions and self excitation in inhomogeneous primary radiation, is derived by integrating the equation (8.32) with respect to the wavelength of the primary spectrum within the range from the short-wave boundary to the K-absorption edge (if the K→L transition is examined) of the element under consideration:

$$I_i = \frac{s}{4\pi R^2}\frac{1}{\lambda_i}\int_{\lambda_0}^{\lambda_K}\frac{\lambda I(\lambda)\vartheta_i(\lambda)d\lambda}{\frac{\mu(\lambda)}{\sin\varphi}+\frac{\mu(\lambda_i)}{\sin\psi}}, \tag{8.70}$$

where $\vartheta_i(\lambda) = \tau_{Ki}(\lambda)\omega_L p_{Li} c_i\left(P_{KL} + \frac{1}{2}\tau_{Li}(\lambda_{Ki})\omega_K c_i L(\lambda)\right)$ is the probability of transformation of the primary photon to a fluorescence one, combining the previously mentioned probabilities for the cascade transitions and self-excitation.

In inhomogeneous primary radiation, the wavelength range between K- and L-absorption edges includes part of this radiation that effectively excites the L-shell without participation of cascade transitions and the self-excitation process (Fig. 8.14).

The effect of the thickness of an aluminium filter, attenuating the long-wave primary radiation of an x-ray tube with a silver anode (calculated in [305]), on the intensity of the individual components of L-fluorescence of antimony is shown in Table 8.2.

Table 8.2 shows that with a decrease of the thickness of the filter from 1 mm to 0 the effect of the cascade transitions and of the self excitation process on the intensity of x-ray fluorescence decreases from ~33% to ~1%.

8.9.3.2. Ionisation of atoms by photo- and Auger electrons

The equation for the intensity, excited by photoelectrons of some element in the conditions of inhomogeneous primary radiation, is derived by integrating equation (8.32) with respect to the wavelength of the primary spectrum:

$$I_i = \frac{s}{4\pi R^2}\frac{1}{\lambda_i}\int_{\lambda_0}^{\lambda_m}\frac{\lambda I(\lambda)\vartheta_i(\lambda)d\lambda}{\frac{\mu(\lambda)}{\sin\varphi}+\frac{\mu(\lambda_i)}{\sin\psi}}, \quad (8.71)$$

where $\vartheta_i(\lambda) = \omega_q p_{qi}\sum_j c_j \sum_m \tau_{mj}(\lambda) n_{mj}(c_i, E)$ is the probability of transformation of a primary photon to a fluorescent one.

The energy of the photoelectrons should be sufficient of ionisation of the atoms of element i. Therefore, the upper boundary λ_m of integration in (8.71) is determined by the value $\lambda_m = \frac{\lambda_{Ki}\lambda_{qJ}}{\lambda_{Ki}+\lambda_{qJ}}$, where λ_{Ki} and λ_{qJ} are the wavelengths of the edges of adsorption of the examined element i and element J supplying the photoelectrons, respectively.

The intensity of element i, excited by the Auger electrons, is determined by the expression:

$$I_i = \frac{s}{4\pi R^2}\frac{1}{\lambda_i}\int_{\lambda_0}^{\lambda_{qj}}\frac{\lambda I(\lambda)\vartheta_i(\lambda)d\lambda}{\frac{\mu(\lambda)}{\sin\varphi}+\frac{\mu(\lambda_i)}{\sin\psi}}, \quad (8.72)$$

where $\vartheta_i(\lambda) = \omega_q p_{qi}\sum_j c_j\tau_j(\lambda)(1-\omega_j)(0.7n_j(c_i, E_{KLL})+0.3n_j(c_i, E_{KLM}))$ is the probability of transformation of the primary photon to the fluorescent radiation photon.

The upper integration limit in equation (8.72) is determined by the wavelength of the absorption edge of the element J which supplies the Auger electrons, because the energy of these electrons is independent of the energy of the primary radiation quantum. The effect of K-electrons on the intensity of x-ray fluorescence of sodium, magnesium, aluminium and sulphur in inhomogeneous primary radiation of an x-ray tube with a palladium anode was investigated in [309] and supplemented in [310] by the investigations of the effect of the L-, photo- and K- Auger electrons on the intensity of sodium. The results of calculations of the contribution of the investigated electrons to the intensity of sodium for the low sodium content of the specimens with different fillers are presented in Fig. 8.15.

Comparison of the results are shown in Fig. 8.15, with the data in Fig. 8.9 for the case of monochromatic primary radiation of the radioactive source ^{109}Cd (the photon energy 22.1 keV) shows that the contribution of photo- and Auger electrons decrease many tens of times and equals units of percent. This is easily explained by the high intensity of the long-wave component of primary bremsstrahlung in direct excitation of the sodium atoms. The decrease of the thickness of the beryllium window on the x-ray

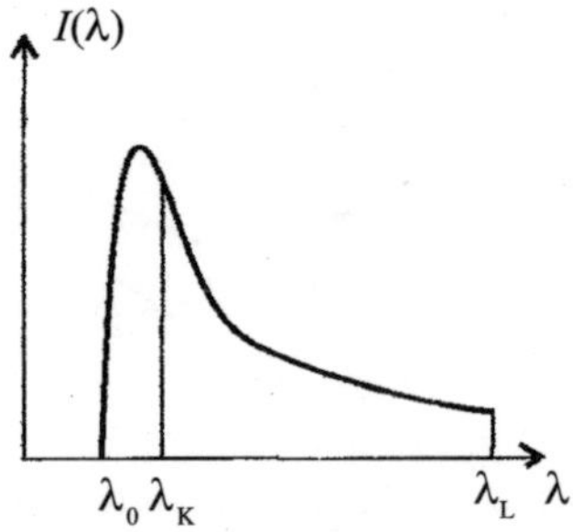

Fig. 8.14. Diagram for evaluating the contribution of cascade transitions to the intensity of x-ray fluorescence in primary bremsstrahlung . The part of the bremsstrahlung spectrum, inducing the cascade transitions and self-excitation, is crosshatched.

Table 8.2. Dependence of the components of x-ray L-fluorescence of antimony (%) on the thickness of the aluminium filter by primary radiation of an x-ray tube with a silver anode (voltage 40 kV, $\varphi = 30°$, $\psi = 90°$)

Filter thickness, mm	1.0	0.5	0.1	0
Photoionisation	66.96	81.35	95.22	99.05
K-L-transitions	28.15	15.89	4.07	0.81
Self-excitation	4.89	2.76	0.71	0.14

tube results in a further decrease of the investigated contribution of the photo- and Auger electrons.

The contribution of the Auger electrons, shown in Fig. 8.15, does not have the jump shown in Fig. 8.9. This is explained by the fact that at the x-ray tube voltage of 25 kV, the primary radiation for the investigated set of the fillers is purely bremsstrahlung. Therefore, with an increase of the atomic number Z_f the bremsstrahlung spectrum, capable of exciting atoms of the filler, becomes more and more restricted and this results in a smooth decrease of the contribution of the Auger electrons down to zero when the wavelength of the absorption edge of the filler is equal to the short-wave boundary of bremsstrahlung.

At a sodium content close to 100%, the contribution of the photoelectrons rapidly decreases because the ionisation cross-section of the K-shell (and, even more so, of the L-shell) of the sodium atoms by the photons with relatively high energies is not large. It should be taken into account that the Auger K-electrons of sodium are not capable of ionising its K-shell.

With the increase of the atomic number of the investigated element the contribution of the photo- and Auger-electrons to the formation of the fluorescent radiation of the element rapidly decreases, and at $Z > 15$ it may be ignored [310].

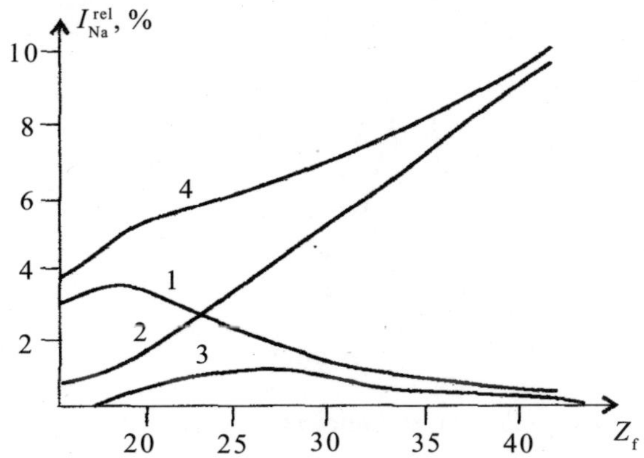

Fig. 8.15. The contribution of photo- and Auger electrons of elements with the atomic number Z_f to the excitation of sodium atoms (composition of specimens 0.1% Na +99.9% filler) in primary radiation of an x-ray tube with a palladium anode (voltage 25 kV, thickness of the beryllium window 300 μm). 1 and 2) the contributions of K- and L-photoelectrons, respectively, 3) the contribution of Auger electrons, 4) the total contribution of the photo- and Auger electrons.

8.9.3.3. Selective excitation effect

The equation for the intensity of x-ray fluorescence of element i, determined by the selective excitation effect of its atoms by the radiation of some element j in polychromatic primary radiation, is derived by integrating equation (8.32) with respect to the wavelength in the range from the short,wave boundary λ_0 of the bremsstrahlung to the absorption edge of the exciting element:

$$I_i = \frac{s}{4\pi R^2}\frac{1}{\lambda_i}\int_{\lambda_0}^{\lambda_{qJ}} \frac{\lambda I(\lambda)\vartheta_i(\lambda)d\lambda}{\dfrac{\mu(\lambda)}{\sin\varphi}+\dfrac{\mu(\lambda_i)}{\sin\psi}}, \tag{8.73}$$

where the probability $\vartheta_i(\lambda)$ of the transformation of the primary photon to a fluorescent one is determined by the dependence:

$$\vartheta_i(\lambda) = P_1 L P_2, \tag{8.74}$$

in which, according to (8.38) and (8.44):

$$P_1 = \tau_{q_1 j}(\lambda)\omega_{q_1 j} c_j,\; P_2 = \tau_{qi}(\lambda_j)\omega_{qi} p_{qik} c_i,$$

$$L = \frac{1}{2}\left(\frac{\sin\varphi}{\mu(\lambda)}\ln\left(1+\frac{\mu(\lambda)}{\mu(\lambda_j)\sin\varphi}\right)+\frac{\sin\psi}{\mu(\lambda_i)}\ln\left(1+\frac{\mu(\lambda_i)}{\mu(\lambda_j)\sin\psi}\right)\right)$$

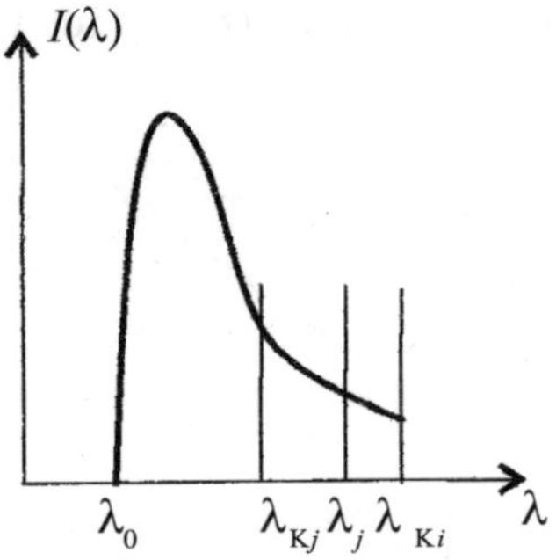

Fig. 8.16. Diagram of the selective excitation effect for inhomogeneous primary radiation.

Examination shows that primary radiation appears in the wavelength range between the absorption edges of the elements *i* and *j*. This radiation directly excites the atoms of the element *i* and does not take any part in the selective excitation process (Fig. 8.16).

Therefore, the contribution of the selective excitation effect in the formation of x-ray fluorescence of the element *i* in the case of inhomogeneous primary radiation decreases in comparison with the case of monochromatic primary radiation. However, the effect remains very strong at a low content of the excited element and a high content of the exciting elements for slightly absorbing matrices. For example, at C_{Fe}=1%, C_{Ni} = 60% and a slightly absorbing filler with primary radiation of the x-ray tube with a tungsten anode (variant close to maximum), selective excitation equals ~90% of the directly excited fluorescence of iron.

The interval between the absorption edges of the elements *i* and *j* increases with an increase of the atomic number of element *j*. Correspondingly, the contribution of the selective excitation effect decreases. At the same time, the wavelength λ_j of the exciting element moves away from the absorption edge of the element *i* and this also reduces the investigated contribution. If L-radiation of some element causes excitation, instead of K-radiation, the contribution of the selective excitation effect decreases by approximately an order of magnitude [326].

The selective excitation processes may include a third order effect (section 8.6.4). The intensity of x-ray fluorescence, formed in inhomogeneous primary radiation as a result of this effect, is determined by the same equation (8.73) in which integration with respect to the wavelength is carried out to the absorption edge of the element, absorbing the primary radiation photons. In addition to this, the probability of transformation $\vartheta_i(\lambda)$ of the primary photon to a fluorescent one in equation (8.73) is determined, for the third order effect, by the equations (8.48) and (8.49). To a first approximation, this effect is maximum when the factors of the chemical composition W_i, determined by equation (8.29) for two exciting elements are approximately identical and equal to ~0.5, and the content of the excited element is low [325].

For a Ni–Cr–Fe system at inhomogeneous primary radiation of an x-ray tube with a palladium anode, working at a voltage of 40 kV, the maximum contribution of this effect to the excitation of x-ray fluorescence of ion is approximately 5%.

In the presence of several exciting elements and of several combinations of these elements, causing third order effects, it is necessary to sum up the equations (8.73) with respect to all possible manifestations of the investigated processes.

8.9.3.4. Perturbing effect. Compensation

The perturbing effect is determined as the total influence of the selective excitation effects and selective absorption of primary radiation. In inhomogeneous primary radiation, using the relationships (8.11) and (8.58), the following equation may be written for the intensity of x-ray fluorescence of the element i taking into account the perturbing effect of element j:

$$I_i = \frac{s}{4\pi R^2}\frac{1}{\lambda_i}\int_{\lambda_0}^{\lambda_{qi}}\frac{\lambda I(\lambda)\vartheta_i(\lambda)}{\mu'(\lambda)+m\mu(\lambda_i)}(1+c_j\beta_{ij})d\lambda, \tag{8.75}$$

where

$$\beta_{ij} = \frac{\int_{\lambda_0}^{\lambda_{qj}}\frac{\lambda I(\lambda)}{\mu(\lambda)+m\mu(\lambda_i)}(\vartheta_i^D(\lambda)-\tau_{qj}(\lambda))d\lambda}{\int_{\lambda_0}^{\lambda_{qi}}\frac{\lambda I(\lambda)\vartheta_i(\lambda)d\lambda}{\mu'(\lambda)+m\mu(\lambda_i)}}. \tag{8.76}$$

The dependence of β_{ij} on the atomic number Z_j of the exciting element for low aluminium contents in element j at primary bremsstrahlung of the x-ray tube with a palladium anode (V = 30 kV, the thickness of the beryllium window 300 μm) and at φ = 50°, ψ=35°, is shown in Fig. 8.17.

The presence of the maximum in Fig. 8.17 is explained at the fact that at Z_j values, close to the atomic number of element i, the selective absorption of primary radiation is very strong. With an increase of Z_j the degree of selective absorption rapidly decreases with a simultaneous increase of the fluorescence yield of the j-th element. These effects are stronger than the effect of movement of the wavelength λ_j from the absorption edge of element i. Therefore, the magnitude of the correction β_{ij} increases, reaching a

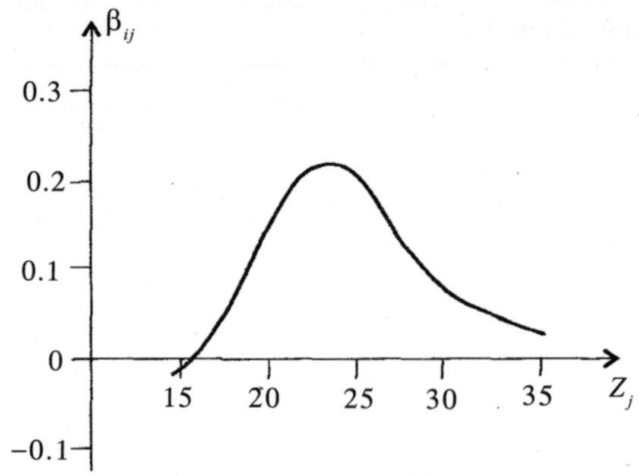

Fig. 8.17. Dependence of correction β_{ij} on the atomic number Z_j of the perturbing element for binary mixtures with a low aluminium content of element j.

maximum at $Z_j = 23–25$. At higher values of Z_j, the selective absorption is almost insignificant and the controlling factor is the increase of the distance between λ_j and the adsorption edge of element i leading to a decrease of the strength of the perturbing effect. This dependence of the perturbing effect on the atomic number of the perturbing element is observed not only for aluminium but also for other elements.

Figure 8.17 also shows that, in certain conditions, the perturbing effect is equal to zero. This means that the total compensation of the selective excitation process by selective absorption of primary radiation, examined in section 8.8 for monochromatic primary radiation, is fully possible also in inhomogeneous primary radiation. This takes place when $\beta_{ij} = 0$, i.e., if the denominator in equation (8.76) is equal to zero:

$$\int_{\lambda_0}^{\lambda_{qj}} \frac{\lambda I(\lambda)}{\mu(\lambda) + m\mu(\lambda_i)} (\vartheta_i^D(\lambda) - \tau_{qi}(\lambda) d\lambda) = 0. \qquad (8.77)$$

At the given spectral composition of radiation of the x-ray tube, the compensation conditions may be fulfilled by changing the value m which characterises of the ratio of the incidence angle of primary and of the selection angle fluorescent radiation.

For the specimens with a low rubidium content and greatly changing content of the perturbing elements (zirconium, niobium and molybdenum), equation (8.77) is used to calculate the selection angles of x-ray fluorescence, corresponding to the compensation conditions. The results of the calculations are compared in Table 8.3 with the experimental data obtained in a spectrometer with different incidence angles of primary radiation and selection angles of fluorescent radiation ($\varphi + \psi = 90°$). The spectrometer

Table 8.3. Comparison of the calculated and experimentally determined compensation angles for radiation of rubidium

$W_{\text{Rb J}}$	Compensation angles for *J*-th perturbing elements					
	Zr		Nb		Mo	
	Calculated	Experiment	Calculated	Experiment	Calculated	Experiment
0.1	53.5	53.0	55.3	59.0	58.5	57.5
0.2	52.3	51.5	54.6	58.5	57.8	59.0
0.3	51.5	52.0	53.2	58.0	56.3	57.0
0.4	50.7	51.5	52.7	56.6	55.2	56.0
0.5	49.9	50.0	51.6	54.0	54.7	55.0

was fitted with an x-ray tube with a tungsten anode working at a voltage of 40 kV.

Of the 16 specimens prepared for the experiments, one specimen did not contain perturbing elements, and the others were divided into three groups containing the oxides of zirconium, niobium and molybdenum. The filler was selected to ensure that the chemical composition factor $W_{\text{RbRb}}=0.1$, and the factor $W_{\text{RbJ}}=C_j\mu_j(\lambda_i)/\mu(\lambda_i)$ (element *J* is one of the perturbing elements) have the values of 0.1, 0.2, 0.3, 0.4 and 0.5. The introduction of the chemical composition factors makes it possible to reduce the number of variables characterising all possible variations of the chemical composition [337]. The selection angle of radiation of thc Kα-linc of rubidium changed from 10° to 80°. A compensation angle at which the measured intensities of the specimens containing and not containing perturbing elements were identical, was determined.

The data in Table 8.3 indicate that the results of the calculations and experimental results are in good agreement. It is important to note the weak dependence of the compensation angle on the content of the perturbing element.

8.10. Special features of the excitation of x-ray fluorescence of elements with low atomic numbers

In section 8.9.3.2 it was noted that at a thickness of the beryllium window of 300 μm, the contribution of photo- and Auger electrons to the intensity of x-ray fluorescence of sodium in inhomogeneous primary radiation of the x-ray tube with the palladium anode may equal several percent. For elements with low atomic numbers (B, C, N, O, F), the role of these electrons is considerably more important because the long-wave radiation, characterised by the highest efficiency in photoionisation of the atoms of these elements, is fully absorbed in the x-ray tube window. A decrease of the thickness of the beryllium window to 75 μm increases the intensity of elements with the low atomic numbers by a factor of 3–4. However, even at a window thickness of 75 μm, the radiation with the wavelength of 12 Å attenuates by approximately a factor of 1000. Figure 8.18 shows schematically the distribution of the K-edge of absorption of carbon in relation to primary radiation for a 75 μm beryllium window. The figure indicates that for this window, the range of the wavelength values from ~10 Å to 43.6 Å does not contain primary radiation.

The processes of formation of x-ray fluorescence of the elements with low atomic numbers have been examined in [338, 339]. The contribution of electrons to the excitation of x-ray fluorescence is minimal when the specimen consists of a single element and the intensity is influenced only by its own K-photoelectrons (probability of ionisation of the K-shell is negligibly small, and the Auger K-electrons have a very low energy). The magnitude of the contributions for the monoelement specimens in primary radiation of the x-ray tube with a rhodium anode and the thickness of the beryllium window of 75 μm was calculated for the voltage of 40 kV using equation (8.71) in which the spectral distribution parameter radiation is given for an earthed cathode (section 3.1.2). The results of the calculations are presented in Table 8.4.

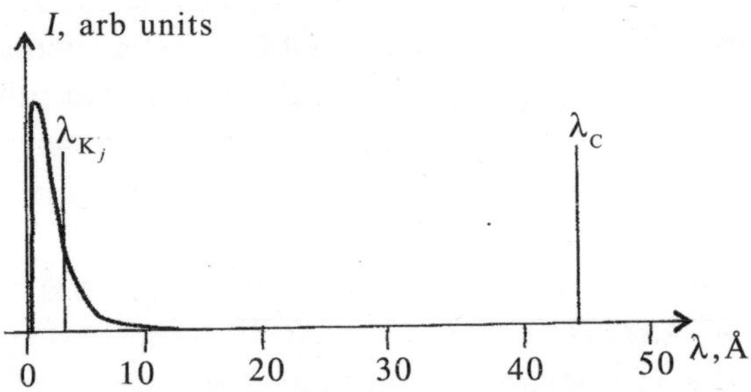

Fig. 8.18. The diagram for the substantiation of the role of photo- and Auger electrons and excitation of x-ray fluorescence of elements with low atomic numbers. The wavelength scale gives the K-edge of absorption of carbon and the K-edge of absorption of some element *j*, the supplier of photo- and Auger electrons.

Table 8.4. Contribution of the K-photoelectrons to excitation of x-ray fluorescence of monoelement specimens with low atomic numbers

Element	C	N	O	F
$I_i^{pe}/I_i + I_i^{pe}$, %	17.7	9.0	4.8	2.5

Table 8.4 shows that the contribution of the photoelectrons to the excitation of fluorescence is relatively small and rapidly decreases with the increase of the atomic number.

However, the contribution of photo- and Auger-electrons to the excitation of fluorescence rapidly increases if the irradiated specimen contains elements with mean and high atomic numbers (for example, some element j whose K-edge of absorption is shown in Fig. 8.18). The atoms of these elements are effectively excited by primary radiation, and the high energy of their photo- and Auger-electrons ensures effective ionisation of the K-shell of the elements with low Z. Table 8.5 shows the components of the total intensity of x-ray fluorescence of carbon at a low carbon content (0.1%) in a silicon and and iron matrix for primary radiation of the x-ray tubes with an earthed cathode (rhodium anode, thickness of the beryllium window 75 μm, voltage 40 kV).

Table 8.5 shows that the contribution of photo- and Auger electrons forms the main part of the x-ray fluorescence of carbon. The individual components of the total contribution greatly change. If the filling matrix contains elements with high atomic numbers, then at a tube voltage of the order of 40–50 kV, the K-shell of these elements does not produce photo- and Auger electrons. On the other hand, it is necessary to take into account M-, N-photoelectrons and L- and M-Auger electrons. The components of the intensity of x-ray fluorescence of carbon, determined by the excitation of the carbon atoms by photoelectrons and Auger electrons of the elements of different fillers, were calculated in [340] and are presented in Fig. 8.19. The composition of the specimens: 0.1%C+99.9% of the element of the filler. All the intensities related to the intensity of carbon directly excited by the primary radiation of the x-ray tube. The selective excitation of the carbon atoms was low and was not considered in the calculations.

Figure 8.19 shows that each of the investigated dependences of the relative intensity of the fluorescence of carbon, determined by the excitation of carbon atoms by M-, N-photoelectrons and by Auger L- and M-electrons, has a maximum, i.e., it is similar to the previously explained dependences

Table 8.5. Components of the intensity of x-ray fluorescence of carbon at a low carbon content in silicon and iron interfering matrices

Composition	X-ray fluorescence components					
	Direct excitation	Selective excitation	K-photoelectrons	L-photoelectrons	Auger electrons	All electrons
0.1 % C + 99.9 % Si	27.6	2.9	22.7	7.2	39.6	69.5
0.1 % C + 99.9 % Fe	25.5	0.2	10.1	51.5	12.6	74.3

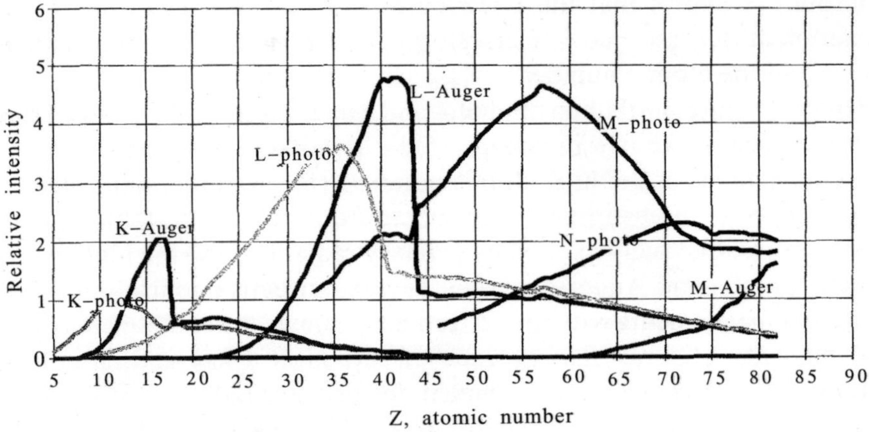

Fig. 8.19. Dependence of the relative intensity of radiation of the Kα-line of iron excited by all types of photoelectrons and Aage-electrons of the filler, on eight atomic number (rhodium anode, d_{Be}=75 μm, V=40 kV).

of the relative intensity of carbon for K-, L-photoelectrons and Auger K-electrons. However, as expected, the maxima of the newly determined dependences for the electrons of the shells found at large distances from the nucleus are shifted to the region of the elements with high Z_f. The figure also shows that for the fillers with the atomic number $Z < 30$, the effect of M- and N-photoelectrons and Auger L- and M-electrons may be disregarded.

The data in Fig. 8.19 lead to a conclusion according to which the composition of the irradiated material strongly influences the value of the contribution of photo- and Auger electrons to the formation of the intensity of x-ray fluorescence of elements with low atomic numbers. The effect of the spectral composition of primary radiation is also important. Therefore, when selecting the optimum conditions for achieving the maximum intensity of x-ray fluorescence of these elements, it should be taken into account that the direct excitation of their atoms by primary photos is not controlling.

The controlling role of photo- and Auger electrons in the excitation of x-ray fluorescence of elements with low atomic numbers has been confirmed by experiments by comparing the calculated and measured intensities of x-ray fluorescence of carbon in a number of its compounds. Experiments were carried out in ARL-9800 x-ray spectrometer with a 3GM x-ray tube, with the beryllium window 75 μm thick. Tube voltage was 40 kV. Carbon compounds were selected as irradiated specimens because of the need to minimise the effect of the particle size on the intensity of x-ray fluorescence. The intensity of fluorescence, related to the intensity of carbon for graphite specimens, is shown in Table 8.6. The table also gives the results of

Table 8.6. Comparison of the experimental and calculated intensities of the Kα-line of carbon in different compounds, related to its intensity for graphite (x-ray tube with a rhodium anode, the thickness of the beryllium window 75 μm. Voltage 40 kV, current 60 mA)

Compound	Experiment	Calculations	
		Taking photo- and Augerelectrons into account	Not taking photo- and Auger electrons into account
Li_2CO_3	0.088±0.004	0.087	0.080
$MgCO_3$	0.050±0.004	0.051	0.035
K_2CO_3	0.074±0.006	0.077	0.044
$K_4Fe(CN)_6$	0.121±0.006	0.120	0.073
$Y_2(CO_3)_3$ $3H_2O$	0.051±0.006	0.054	0.019
$La(CH_2COO)_3$	0.274±0.006	0.286	0.093
$Tm_2(CO_3)_3$ $3H_2O$	0.063±0.006	0.072	0.015
$Pb(CH_3COO)_2$	0.286±0.006	0.298	0.031

calculations, carried out without taking the effect of photo- and Auger-electrons into account.

Table 8.6 shows that the experimental data are in good agreement with the results of calculations which take into account the effect of photo- and Auger electrons, and greatly differ from the intensities calculated without taking this effect into account.

In the conditions with low efficiency of direct ionisation by primary radiation of the atoms of the elements with low atomic numbers, it is necessary to pay special attention to the effect of selective excitation, which may be manifested in three variants. It is the ionisation of the atoms by K- and L-radiation of the exciting elements, and also by L-radiation of these elements, formed as a result of intra-atomic cascade transitions. The cascade transitions often transform the primary photons of the element-filler to photons with low energy which are far more effective for excitation of x-ray fluorescence of the elements with low atomic numbers Z. For example, the energy of L-radiation of iron (~0.7 keV) is closer to the binding energy of K-electrons in the atoms of carbon, nitrogen, oxygen and fluorine than the energy of K-radiation of iron (~6.4 keV). Of course, the coefficient of the K–L-transformation in cascade transitions, determined by the yield of x-ray L-fluorescence ω_L, is small (for iron $\omega_L \approx 0.6\%$). With the increase of the atomic number of the excited element, the energy of its K-edge of

Table 8.7. Contribution of the process of selective excitation to the formation of x-ray fluorescence of elements with low atomic numbers at a content of the elements of 0.1% in several interfering matrices

Matrix	Contribution of individual processes (%)					
	Direct excitation	Selective excitation				Excitation by photo- and Auger electrons
		K-	L-	K-L-	Σ	
Carbon						
Fe	21	0.3	4	0.4	4.7	74.3
Ni	20	0.2	3.9	0.4	4.5	75.5
Zn	21	0.1	3.3	0.2	3.6	75.4
Nitrogen						
Fe	43	0.7	8.7	0.8	10.2	46.8
Ni	42	0.5	9.5	0.9	10.9	47.1
Zn	47	0.2	8.5	0.4	9.1	43.9
Oxygen						
Fe	58	1.2	14	1.3	16.5	25.5
Ni	58	0.9	15	1.5	17.4	24.6
Zn	64	0.4	14	0.6	15	21.0
Fluorine						
Fe	65	1.8	19	1.8	22.6	12.4
Ni	64	1.3	20	2.1	23.4	12.6
Zn	70	0.6	18	0.9	19.5	10.5

absorption approaches the energy of L-radiation of ıron, and the role of L-selective excitation in the formation of K-fluorescence of the investigated elements should increase.

The contribution of each of the three variants of this effect to the intensity of x-ray fluorescence of carbon, nitrogen, oxygen and fluorine at a low content of these elements in the iron matrix (99.9% of iron and 0.1% of the investigated elements) is reflected in Table 8.7. The primary radiation was generated by an x-ray tube with a tungsten anode at a voltage of 40 kV.

Table 8.7 shows that of the three variants of the selective effect, the more significant variant is the selective excitation of the atoms by the L-radiation of iron. The contribution of K–L-cascade transitions to L-radiation of the filler is not so large. The total contribution of the effect of selective excitation rapidly increases with an increase of the atomic number of the element A (from units of percent for carbon to 22.6% for fluorine). The role of the effect of selective excitation and the ratio of the contributions of the individual components of excitation depend on the spectral composition of primary radiation.

Conclusions

In conclusions, we examine several examples of the application of x-radiation in various areas of science and production.

X-radiation found applications almost immediately after its discovery. For example, already in 1896, the inventor of the radio A.S. Popov used x-radiation for detecting small shot in the body of an injured man. At present, this type of radiation is used widely in medicine for diagnostics and therapy purposes. In particular, the position and spatial shape of heterogeneities in tissue (x-ray tomography) are determined with high accuracy. Medical tomographs are complicated and efficient systems for x-ray radiation in several directions for the accurate diagnostics of the state of internal organs.

Similar problems are also solved by x-ray defectoscopy which determines the shape and position of heterogeneous inclusions in natural objects and industrial components. In addition to the simplest irradiation methods, defectoscopy uses also tomographs designed on the basis of specific practical tasks. For example, in aircraft construction, in detection of defects in the wings when radiation cannot be used, the problem is solved by using the dependence of the intensity of Compton scattering on the atomic number of the scattering objects.

The phenomenon of diffraction of x-rays on ordered structures represented the basis of development of x-ray diffraction analysis which is an important method in science and technology. This method is used widely for the identification of particles and determination of the concentration in powder materials (phase analysis), for the determination of the atomic and molecular structure of single crystals, organic compounds and other objects. X-ray diffractometers are used for solving problems in crystallography, organic and inorganic chemistry, for the control of the structural characteristics of industrial components, etc.

The dependence of the wavelength of characteristic radiation on the atomic number of the emitting element (Moseley law) is the basis of x-ray spectrum analysis of substances and materials. The currently available equipment and theory of the dependence of the intensity of x-ray radiation on the elemental composition of the analysed material ensure high accuracy of the determination of the content of elements so that it has been possible in a number of cases to replace the time-consuming and laborious chemical methods of analysis by the express x-ray spectrum fluorescence method. X-ray fluorescence spectrometers determine in several minutes the content of 20-30 elements in materials with complicated composition. The method is irreplaceable in the express control of technological processes and quality of industrial production.

The intensity of x-ray fluorescence may also be used as an indicator of the thickness of films, layers and coatings. The nondestructive x-ray fluorescence inspection of thin-layer components is used in various areas of industry (metallurgy, electroplating, engineering, electronics, etc). The range of controlled thicknesses is not overlapped by any other analysis methods. The method itself is highly universal and can be used for greatly differing materials forming the layer and coating. In addition to thickness, the method can be used for the rapid inspection of the elemental composition of thin-layer components.

The fact that the electron beam can be sharply focused is the basis of x-ray microprobe analysis of primary spectra which is an important tool in investigating substances. The method can be used for the determination of the elemental composition of micron volumes of the investigated materials. The methods is almost irreplaceable in investigating the composition of microparticles, in the determination of the distribution of elements of the area or volume of the objects, in the investigations of contacts between different inclusions and in investigating boundary, including subsurface layers, etc.

Fluorescence microanalysis represents a serious competitor to microprobe analysis of primary spectra. As a result of the rapid development of x-ray focusing optics, it has become possible to determine the elemental composition in micron volumes at relatively large depths in the investigated object. Investigations can be carried out without any special preparation of samples and without evacuation of the volume.

The dependence of the structure of x-ray absorption spectra and also of the shape and position of certain x-ray lines on the state of matter enables us to determine using x-ray spectroscopy the type of chemical bond, magnetic properties, electrical conductivity and other characteristics of the investigated materials. Together with the methods of electron and Auger spectroscopy, this widely used method is a powerful tool for investigating matter.

The application of the effect of total external reflection of x-ray radiation from efficiently produced surfaces has made it possible to lower greatly the limit of detection of a number of elements in the subsurface layers and microvolumes of material deposited on such as surfaces, greatly widening the areas of possible application of x-ray spectrum investigations.

There is also a number of other applications of x-radiation (shadow microscopy, density measurements, sterilisation of products, etc).

A large number of x-ray instrumentmaking companies have been formed for the practical realisation of these applications of x-ray radiation. A number of important national and international firms and companies produce, adapt for specific applications and introduce x-ray equipment and appropriate mathematical apparatus in various branches of science and production.

References

1. Rontgen W.C. Uber eine neue Art von Strahlen, 1, Sitzunger. Med.phys. Ges., 1895, p.137.
2. Compton A., Allison S., X-ray beams. Theory and experiment, [Russian translation], Moscow–Leningrad, GITTL, 1941.
3. Sommerfeld A., Structure of atom and spectra, [Russian translation], ?oscow–Leningrad, GITTL, 1956.
4. Geitler W., Quantum theory of radiation, [Russian translation], Moscow, IL, 1956.
5. Korsunskii M.I., Physics of X-rays, Moscow, ONTI, 1936
6. Borovskii I.B., Physical fundamentals of X-ray spectrum studies, Moscow University, 1956.
7. Blokhin M.A., Physics of X-rays, Moscow, GITTL, 1957.
8. Alikhanov A.I., X-ray optics, Moscow, GTTI, 1933.
9. Wilson A., X-ray optics, [Russian translation], Moscow, IL, 1951.
10. Aristov V.V., Erko R.K., X-ray optics, Moscow, Nauka, 1991
11. Wooster W., Diffusion scattering of x-ray in crystals, [Russian translation], Moscow, IL 1963.
12. Iveronova V.I., Theory of scattering of X-rays, Moscow State University, Moscow, 1972.
13. Amus'ya M.Ya., Bremsstrahlung, Energoatomizdat, 1990.
14. Bertin E.P., Principles and practice of X-ray spectrometric analysis, N.Y.: Plenum Press, 1970.
15. Jenkins R., DeVries J., Practical *X-Ray Spectrometry*, Eindhoven, MacMillan, 1972.
16. Azaroff L.(editor), *X-Ray Spectrometry*, McGraw Hill, Berks, Maidenhead, 1974.
17. Bertin E.P., Introduction in X-ray spectrometric analysis, N.Y., Plenum Press, 1978.
18. Reed S., Electron probe microanalysis [Russian translation], Moscow, Mir, 1979.
19. Heinrich K.F.J., X-ray microanalysis, N.Y. Van Nostrand Reinhold Company, 1981.
20. Scott W.D., Love G., Quantitative X-ray probe microanalysis, [Russian translation], Moscow, Mir, 1986.
21. Pavlova L.A., Paradina L.F., X-ray spectrum microanalysis and application in mineralogy, Yakutsk, 1990..
22. Laputina I.P., Microanalysis in mineralogy, Moscow, Nauka, 1991.
23. Pavlova L.A., Beloserova O.Yu., Paradina L.D., Suvorova L.F., X-ray spectrum electron-probe microanalysis of natural and geological objects, Irkutsk, 1998.
24. Warren B.E., X-Ray diffraction, Dover Publications, 1990.
25. Muller R.O., Spektrochemische Analysen mit Rontgenfluoreszenz, Munchen, R. Oldenbourg, 1967.
26. Losev N.F., Quantitative X-ray spectrum fluorescence analysis, Nauka, Moscow, 1969.
27. Losev N.F., Smagunova A.M., Fundamentals of X-ray spectrum fluorescence analysis, Khimiya, Moscow, 1982.
28. Afonin, V.P., et al., X-ray fluorescence analysis, Nauka, Novosibirsk, 1991.
29. Afonin V.P., Gunicheva T.N., X-ray spectrum fluorescence analysis of rocks and

minerals, Nauka, Novosibirsk, 1984.
30. Afonin V.P., Gunicheva T.N. and Piskunova L.F., X-ray fluorescence silicate analysis, Nauka, Novosibirsk, 1984.
31. Bakhtiarov A.V., X-ray spectrum fluorescence analysis in geology and geochemistry, Nedra, Moscow, 1985.
32. Revenko A.G., X-ray spectrum fluorescence analysis of natural materials, Nauka, Novosibirsk, 1994.
33. Borkhodoev V.Ya. X-ray spectrum fluorescence analysis of rock by the method of fundamental parameters, Magadan, 1999.
34. Rutherford E., *Phil. Mag.*, 1911, ser.6, v.21, p.669–688.
35. Chadwick J., Bieler E.S., *Phil. Mag.*, 1921, v.42, No.252, p.923–940.
36. Bohr N., *Phil. Mag.*, 1913, v.26, p.1–25, 476–502.
37. Stern O., *Zeits. f. Phys.*, 1921, Bd.7, S. 249–253.
38. Schrödinger E., *Ann. d. Phys.*, 1926, v.80, p.437–491.
39. Hartree D.R., Calculations of atomic structures [Russian translation], Moscow, 1960.
40. Moseley H.D.J., *Phil. Mag.*,1914, v.27, No.160, p.703–713.
41. Blokhin M.A., Shveitser I.G., X-ray spectrum handbook, Nauka, Moscow, 1982.
42. Coster D., Kronig R.L., *Physica*, 1935, v.2, p.13–24.
43. Schreiber T.P., *X-Ray Spectrometry*, 1982, v.11, No.2, p.42–45.
44. Marenkov O.S., et al., Relative intensities of fluorescence L-series lines of X-ray spectra. Equipment and methods for X-ray analysis, Leningrad, Mashinostroenie, 1982, No. 27, 132–135.
45. Raghavaiah C.V., et al., *X-Ray Spectrometry*, 1990, v.19, No.1, p.23–26.
46. Smirnov L.A., *Z. Tekh. Fiz.*, 1974, v.44, No.5, p.1110–1113.
47. Wentzel G., *Zeits. f. Phys.*, 1927, Bd.43, S.524–530.
48. Burhop E.H.S., *J. Phys. Rad.*, 1955, v.16, p.625–629.
49. Byrne J., Howarth N., *J. Phys.*, 1970, B3, p.280–291.
50. Hubbell J.H., Bibliography and current status of K, L, and higher shell fluorescence yields for computations of photon energy–absorption coefficients. NISTIR 89-4144. US Department of Commerce. NISTIR. Center of Radiation Research, Gaithersburg. MD 20899, 1989.
51. Bambynek W., New evaluation of K-shell fluorescence yields, Proc. X-Ray and Inter-Shell Processes in Atoms, Molecules and Solids. Leipzig, 20–24 August 1984.
52. Cohen D.D., *Nucl. Instrum. Methods Phys. Res.*, 1987, B22, p.55–58.
53. Bethe H.A., *Ann. Physik*, 1930, Bd.5. p.325–400.
54. Hutchins G.A., Electron probe microanalysis. In: Characterisation of Solid Surfaces, ed. by P.F. Kane and G.B. Larrabee, New York Plenum Press, 1974, p.441–481.
55. Merlet C., *Microbeam Analysis*, 1995, v.4, p.239–253.
56. Hoeft H., Schwaab P., *X-Ray Spectrometry*, 1988, v.17, N5, p.201–208.
57. Widdington R., *Proc.Roy.Soc.*, Ser.A, Math. and Phys. Sci., 1912, v.86,No.588, p.360–370.
58. Joy D.C., Luo S., An empirical stopping power relationship for low-energy electrons, Scanning, 1989, v.11, p.176–180.
59. Bishop H.E., Some electron backscattering measurements for solid targets, Optiques des rayons X et microanalyse, Paris, Hermann, 1966, p.153–158.
60. Dupony G., Perrier F., Verdier G., Arnal F., Transmission d'electrons monocinetiques a travels des fenilles metalliques minees, C. a. Acad. Set., 1964, v.258, p.3655–3660.
61. Merlet C., *X-Ray Spectrometry*, 1992, v.21, N.5, p.229–238.

62. Kirkpatrick P., Hare D.G., *Phys.Rev.,* 1934, v.61, p.254–259.
63. Castaing R., Descamps J., *J.Phys.Rad.*, 1955, v.16, N.4, p.304–317.
64. Sewell D., Love G., Scott V.D., *J.Phys. D, Appl. Phys.*, 1985, v.18, N7, p.1233–1243.
65. Pouchou J.-L.,Pishoir E., Quantitative analysis of homogenous or stratified microvolumes applying the model PAP. In: Electron Probe Quantisation, Ed. by K.F.J. Heinrich and D.E. Newbury. N.Y.: Plenum Press, 1991, p.31–75.
66. Bastin G.F., Dijkstra J.M., Heijligers H.J.M., *X-Ray Spectrometry*,1998, v.27, p.3–10.
67. Lavrent'ev Yu.G., Korolyuk V.N., Usova L.V., *Zh. Anal. Khimii*, 2004, v.59, No.7, p.678–696.
68. Philibert J. A method for calculating the absorption corrections in electron probe microanalysis, X-Ray Optics and X-Ray Microanalysis. New York, Academic Press, 1963, p.379–392.
69. Finkel'shtein A.L., Pavlova T.O., *Zavod. Lab.*, 1996, v.62, No.12, p.16–20.
70. Finkel'shtein A.L., et al., *Zavod. Lab.*, 1981, v.47, No.11, p.28–31.
71. Green M., *Proc. Phys. Soc.*, 1963, v.82, p.204–215.
72. Murata K., Matsukawa T., Shimizu R., *Jap. J. Appl. Phys.*, 1972, v.10, No.6, p.678–686.
73. Afonin V.P., Lebed', Monte Carlo method in X-ray spectrum microanalysis, Nauka, Novosibirsk, 1977.
74. Staub P.F., Intri X., *X-Ray Spectrometry*, 1998, v.27, No.1, p.43–57.
75. Staub P.F., Jonnard P., Vergand F., et al., *X-Ray Spectrometry*, 1998, v.27, No.1, p.58–66.
76. Moseley H.G.J., Darwin C.G., *Phil. Mag.*, 1913, v.26, p.210–232.
77. Duane W., Hunt F.L., *Phys. Rev.*, 1915, v.6, p.166–171.
78. Bearden J.A., Johnson F.T, Watts H.M., *Phys.Rev.*, 1951,v.81, issue 1,p.70 –72.
79. Ulrey C.T., *Phys. Rev.*, 1918, v.11, No.5, p.401–410.
80. Muller A., *Phys. Zeits.*, 1918, v.19, No.22, S.489–490.
81. Kulenkampff H., *Ann. der Physik*, 1922, N.69, p.548–597.
82. Kramers H.A., *Phil. Mag.*, 1923, v.46, N.275, p.836–871.
83. Tertian R., Broll N., *X-Ray Spectrometry*, 1984, v.13, No.3, p.134–141.
84. Broll N., *X-Ray Spectrometry*, 1986, v.15, No.4, p.271–285.
85. Ebel H., Ebel M.F., Wernisch J., Phoen E., Wiederschwinger H., *X-Ray Spectrometry*, 1989, v.18, No.3, p.89–100.
86. Statham P.J., *X-Ray Spectrometry,* 1976, v.5, No.3, p.154–167.
87. Keith H.D., Loomis T.C., *X-Ray Spectrometry*, 1976, v.5, p.93–103.
88. Loomis T.C., Keith H.D., *X-Ray Spectrometry*, 1976, v.5, p.104–114.
89. Kulenkampff H., *Ann. d. Phys,* 1928, Bd.87, No.21, S.597–637.
90. Nicholas W.W., *Bureau of Standards Journal of Research*, 1929, v.2, No.5, p.837–870.
91. Sommerfeld A., *Ann. d. Phys.*, F.5, 1931, v.11, No.3, S.257–330.
92. Scherzer O., *Ann. d. Phys.*, F.5, 1928, v.13, No.2, p.137–160.
93. Honerjager R., *Ann. d. Phys.*, F.5, 1940, Bd.38, No.1, p.33–45.
94. Kirkpatrick P., Wiedman L., *Phys.Rev.*, 1945, v.67, p.321–339.
95. Kulenkampff H., *Phys. Zs.*, 1929, Bd.30, No.17, S.513–515.
96. Dasannacharya B., Polarization of X-ray from thin aluminium anti-cathodes: [Abstr.pap.presented at Chicago Meeting of American Physical Society,Nov.29–30,1929], *Phys.Rev.*,1930, v.35, No.1, p.129.
97. Piston D.S., *Phys.Rev.*, 1936, v.49, p.275–279.
98. Boardman B.F.,The polarization of X-ray from thin targets: [Abstr. pap. presented

at the Pasadena, California, Meet. of American Physical Society, June18–20,1941], *Phys.Rev.*,1941, v.60,No.2,p.163.

99. Kirkpatrick P., *Phys.Rev.*, 1923, v.22, p.226–232.
100. Wagner E., Ott P., *Ann. d. Phys.*, 1928, v.85, p.425–470.
101. Pavlinsky G.V., Losev N.F., *Zavod. Lab.*, 1963, v.29, No.9, p.1067–1070.
102. Castellano E.E., Alvarez A.G., Andrieu J.M., *X-Ray Spectrometry*, 1973, v.2, No.4, p.159–163.
103. Pella P.A., Feng L., Small J.A., *X-Ray Spectrometry*, 1985, v.14, No.3, p.125–135.
104. Pella P.A., Feng L., Small J.A., *X-Ray Spectrometry*, 1991, v.20, p.109–111.
105. Afonin V.P., et al., *Zavod. Lab.*, 1968, v.34, No.2, p.169–172.
106. Pavlinsky G.V., Portnoy A.Yu., *Radiation Physics and Chemistry*, 2001, v.62, No.2–3, p.207–213.
107. Pavlinsky G.V., Portnoy A.Yu., *X-Ray Spectrometry*, 2002, v.31, No.3, p.247–251.
108. Komar E.G., Fundamentals of acceleration methods, Atomizdat, Moscow, 1975.
109. Bohr N., Passage of atomic particles through matter [Russian translation], IL, Moscow, 1950.
110. Kolyada V.M., Zaichenko A.K., Dmitrenko R.V., X-ray spectrum analysis with ion excitation, Atomizdat, Moscow, 1978.
111. Garcia J.D., Fortner R.J., Kavanagh T.M., *Rev. Mod. Phys.*, 1973, v.45, N.2, p.111–113.
112. Basbas G., Brandt W., Laubert R., *Phys.Rev.*, 1973, v.A7, p.983–1001.
113. Lapicki G., Goldstein M., Brandt W., *Phys. Rev.* A, 1981, v.23, issue 5, p.2727 – 2729.
114. Lapicky G., *X-Ray Spectrometry*, 2005, v.34, No.4, p.269–278.
115. Lindhard J., Scharff M., *Phys. Rev.*, 1961, v.124, p.128–130.
116. Starodubtsev S.V., Romanov A.M., Passage of charged particles through matter, Acad. Sci. Uzbekh SSR, Tashkent, 1962.
117. Sterk A.A., Advances in X-Ray Analysis, 1965, v.8, p.189–197.
118. Merzbacher E., Lewis H.W., X-ray production by heavy charged particles, In: Handbuch der Physik, Springer Verlag, Berlin, 1958, v.34, p.166–192.
119. Ishii K., Yamazaki H, Matsuyama S., Galster W., Satoh T., Budnar M., *X-Ray Spectrometry*, 2005, v.34, No.5, p.363–365.
120. Ogier W.T., Carlson R.D., Knoche J., *Phys. Rev.*, 1966, v.142, No.1, p.50–52.
121. Eretenko A.N., X-radiation excited by electrons and protons in solids, Dissertation, Rostov-on-Don, 1980.
122. Folkmann F., Gaarde C., Huus T., Kemp K., *Nucl. Instrum. and Methods*, 1974, v.116, p.487–499.
123. Yakubovich A.L., Zaitsev E.I., Przhiyalgovskii S.M., Nuclear physical methods of analysis of mineral resources, Atomizdat, Moscow, 1973.
124. Plotnikov R.I., Pshenichnyi G.A., Fluorescence X-ray radiometric analysis, Atomizdat, Moscow, 1973.
125. Mamikonyan S.V., Equipment and methods of fluorescence radiometric analysis, Atomizdat, Moscow, 1976.
126. Kulipanov G.N., Skrinskii A.N., *Usp. Fiz. Nauk*, 1977, v.122, No.3, p.369–418.
127. Winick H., Doniach S., Synchrotron radiation research, Plenum Press, New York, 1980.
128. Sonntag B., Fresenius Z., *Anal. Chem.*, 1986, Bd.324, S.786–792.
129. Als-Nielsen J., McMorrow D., Elements of modern X-ray physics, John Wiley & Sons, Chichester:, 2001.
130. Baryshev V.B., et al., *Zh. Anal. Khim.*, 1986, v.41, No.3, p.389–401.
131. Jones K.W., Barry M.G., *Anal. Chem.*, 1989, v.61, No.5, p.341A–358A.

132. Ice G.E., *X-Ray Spectrometry*, 1997, v.26, No.6, p.315–325.
133. PerryM.D., Mourou G., *Science*, 1994, v.264, p.917–924.
134. Andreev N.E., Gorbunov L.M., *Usp. Fiz. Nauk*, 1999, v.169, No.1, p.53–58.
135. Rousse A. et al., *Phys. Rev.*, 1994, v.E50, p.2200–2207.
136. Andreev A.A., Yashin V.E., Charukhev A.V., *Usp. Fiz. Nauk*, 1999, v.169, No.1, p.72–78.
137. Kmetec J.D., et al., *Phys. Rev. Lett.*, 1992, v.68, p.1527–1530.
138. Slabko V.V., *Sorosovskii Obraz. Zh.*, 1997, N.1, p.79–85.
139. Documented present to the American Physical Society by an expert group on scientific and technical aspects of beam weapons, *Usp. Fiz. Nauk*, 1988, v.155, p.661–679.
140. Roscn M.D., et al., *Phys. Rev. Lett.*,1985, v.54, p.106–109.
141. Matthews D.L., et al., *Phys. Rev. Lett.*, 1985, v.54, p.110–113.
142. Suckewer S., et al., *Phys. Rev. Lett.*,1985, v.55, p.1753–1756.
143. Mac Gowan B.J., et al., *Fluids Plasma Phys.*, 1992, v.4, No.7, Pt.2, p.2326–2338.
144. Gasparyan P.D., Starikov F.A., Starostin A.N., *Usp. Fiz. Nauk*, 1998, v.168, No.8, p.843–875.
145. Elton R.C., X-ray lasers, Boston: Academic Press, 1990.
146. Arkhipkin V.G., Popov A.K., Nonlinear converters of light in gases, Nauka, Novosibirsk, 1987.
147. Hogan M.J., et al., *Phys. Rev. Lett.*, 1998, v.81, p.4867–4870.
148. Flegel L., Rossbach J., Cern Courier, 2000, v.40, No.6, p.26–28.
149. Stobbe M., *Ann. d. Phys.* F.5, 1930, Bd.7, No.6, p.661–715.
150. Einstein A., Emission and absorption of radiation in quantum theory, In: Studies of kinetic theory, radiation theory and fundamentals of quantum mechanics [Russian translation], vol. 3, Moscow, 1996
151. Robinson H.R., *Proc. Roy. Soc., Ser.A. Math and Phys. Scien.*, 1923, v.104, No.727, p.455–479.
152. Victoreen J.A., *J. Appl. Phys.*, 1949, v.20, p.1141–1147.
153. Viegele W.J., Atomic Data, 1973, v.5, No.1, p.51–111.
154. Scofield J.H., Theoretical photoionisation cross-section from 1 to 1500 keV, Lawrence Livermore Laboratory, 1973, Report UCRL-51326.
155. Chantler C.T., *J. Phys. Chem. Ref. Data*, v.24, p.71 –643.
156. Cullen D.E., Hubbell J.H., Kissel L., EPDL97: The evaluated photon data library, Lawrence Livermore National Laboratory, 1997, Report UCRL–50400, v.6, rev 5.
157. Storm E., Israel H.I., Nucl. Data Tabl., 1970, v.A7, No.6, p.565 – 681.
158. Vedrinskii R.V., Geguzin I.I., X-ray absorption spectra of solids, Energoatomizdat, Moscow, 1991.
159. Vainshtein E.E., X-ray spectra of atoms in molecules of chemical compounds and in alloys USSR Academy of Sciences, Moscow, 1950.
160. Parratt L.G., *Phys. Rev.* 1939, v.56, p.295–297.
161. Sukhorukov V.L., et al., *Optika i Spektroskopiya*, 1983, v.55, No.3, p.229–233.
162. Sukhorukov V.L., et al., *Optika i Spektroskopiya*, 1979, v.47, No.2, .407–409.
163. Mazalov L.N., et al., X-ray spectra of molecules, Nauka, Novosibirsk, 1977.
164. Mazalov L.N., et al., X–ray spectra and chemical bonds, Nauka, Novosibirsk, 1982.
165. Landau L.D., Lifshits E.M., Quantum mechanics, Nauka, Moscow, 1974.
166. Hubbell J.H., McMaster W.H., Del Grande N.K., Mallett J.H., X-ray cross-section and attenuation coefficients, International Tables for Crystallography, 1974, v.4, ed. J.A. Ibers and W.C. Hamilton (Birmingham: Kynoch), p.47–70.
167. Saloman E.B., Hubbell J.H., Scofield J.H., X-ray attenuation cross section for energy 100 eV to 100 keV and elements Z = 1 to Z = 92, Atomic Data and Nucl.

Data Tables, 1988, v.38, p.1–197.

168. Hubbell J.H., Seltzer S.M., Tables of x-ray mass attenuation coefficients and mass energy – absorption coefficients 1 keV to 20 MeV for elements Z = 1 to 92 and 48 additional substances of dosimetric interest, NISTIR 5632, 1995.
169. Henke B.L., Lee P., Tanaka T.J., Shimabukuro R.L., Fujikawa B.K., Low-energy x-ray interaction coefficients: photoabsorption, scattering, and reflection, E = 100–2000 eV, Z = 1–94, Atomic Data and Nucl. Data Tables, 1982, v.27, p.1–144.
170. Henke B.L., Gullikson E.M., Davis J.C., X-ray interaction: photoabsorption, scattering, transmission and reflection at E = 50–30000 eV, Z = 1–92, Atomic Data and Nucl. Data Tables, 1993, v.54, p.181–342.
171. Marenkov O.S., Komyak N.I., Photon coefficients of interaction in X-ray radiometric analysis, a handbook, Energoatomizdat, Leningrad, 1988.
172. Leroux J., *Adv. X-Ray Anal.*, 1962, v.5, p.153–160.
173. Thinh T.P., Leroux J., *X-Ray Spectrometry*, 1979, v.9, No.2, p.85–91.
174. Heinrich K.F.J., Mass absorption coefficients for electron probe microanalysis, In: 11th International Congress on X-Ray Optics and Microanalysis 1986, p.67–119.
175. Tables and formulas for X-ray spectrum analysis: procedure recommendations, compiled by O.S. Marenkov, No. 3, Mashinostroenie, Moscow, 1982.
176. Finkel'shtein A.L., Farkov P.M., *Analitika i Kontrol'*, 2002, v.6, .4, p.377–382.
177. Heckel J., Ryon R.W., Polarized beam X-ray fluorescence analysis, In: Handbook of *X-Ray Spectrometry*, 2001, p.603–630.
178. Thomson J.J., The conduction of electricity through gases, London: Cambridge University Press, 1906.
179. Strutt J.W. (Lord Rayleigh), *Phil. Mag.* (4th ser), 1871, v.41, p.107–274.
180. Strutt J.W. (Lord Rayleigh), *Phil. Mag.* (5th ser), 1881, v.12, p.81–101.
181. Compton A.N., *Phys.Rev.*, 1923, v.21, p.483–502.
182. Klein O., Nishina Y., *Z. Physik*, 1929, v.52, p.853–868.
183. Hubbell J.H., Viegele W.J., Briggs E.A., Brown R.T., Cromer D.T.. Howerton R.J., *J. Phys. Chem. Ref. Data*, 1975, v.4, No.3, p.471–538.
184. Brown R.T., *Phys. Rev.*, 1970, A1, p.1342–1347; A2, p.614–620.
185. Brown R.T., *Phys. Rev.*, 1974, A10, p.438–439.
186. Cromer D.T., *Acta Crystallogr.*, 1968, A24, p.321–324.
187. Hubbell J.H., Overbo I., *J. Phys. Chem. Ref. Data*, 1979, v.8, No.1, p.69–105.
188. Cromer D.T., Weber J.T., Atomic scattering factors for X-rays, International Tables for X-Ray Crystallography, 1974, v.4 (Birmingham: Kynoch), p.71–147.
189. Overbo I., *Nuove Cimento*, 1977, B40, p.330–338.
190. Overbo I., *Phys. Scr.*, 1978, v.17, p.547–548.
191. Tables and formulas for X-ray spectrum analysis: procedure recommendations, compiled by O.S. Marenkov, No. 3, Mashinostroenie, Leningrad, 1982.
192. Bakhtiarov A.V., Pshenichnyi G.A., Formulas for approximate calculation of differential scattering cross-sections of low-energy x-radiation, In: Equipment and methods for X-ray analysis, Mashinostroenie, Leningrad, 1973, No. 12, p. 68–72.
193. Kissel L., Pratt R.H., Roy S.C., *Phys. Rev.*, 1980, A22, p.1970–2004.
194. Pratt R.H., Kissel L., Bergstrom P.M., New relativistic S-matrix results for scattering – beyond the usual anomalous factors / beyond impulse approximation, Resonant Anomalous X-Ray Scattering: Theory and Applications, ed. G. Materlik, C.J.Sparks and K. Fischer (Amsterdam: Elsevier), 1994, p.9–33.
195. Kissel L., *Nucl. Instrum. Methods*, 1995, B99, p.144–147.
196. Cromer D.T., Mann J.B., *J. Chem. Phys.*, 1967, v.47, p.1892–1893.
197. Cromer D.T., *J. Chem. Phys.*, 1969, v.50, p.4857–4859.
198. Ribberfors R., Berggren K.-F., *Phys. Rev.*, 1982, A26, p.3325–3333.

199. Kahane S., Relativistic Dirac–Hartree–Fock photon incoherent scattering functions, At. Data Nucl. Data Tables, 1998, v.68, p.323–347.
200. Wheeler J.A., *Phys. Rev.*, 1937, v.52, p.1107–1122.
201. Heisenberg W., *Z. Phys.*,1943, Bd.120, S.513–538.
202. Heisenberg W., *Z. Phys.*,1943, Bd.120, S.673–702.
203. Rodberg L.S., Thaler R.M., Chapter 9, The S Matrix and K Matrix, Introduction to the Quantum Theory of Scattering (AcademicPress, New York), 1967, p.227–248.
204. Suric T., Bergstrom P.M., Pisk K. and Pratt R.H., *Phys. Rev. Lett.*, 1991, v.67, p.189–192.
205. Suric T., *Nucl. Instr. Meth.*, 1992, A314, p.240–243.
206. Bergstrom P.M., *Nucl. Instr. Meth.*, 1993, B79, p.240–243.
207. Bergstrom P.M., Suric T., Pisk K., Pratt R.H., *Phys. Rev.*, 1993, A48, p.1134–1162.
208. Bergstrom P.M., *Nucl. Instr. Meth.*, 1995, B99, p.148–151.
209. Bergstrom P.M., Pratt R.H., *Radiat. Phys. Chem.*, 1997, v.50, p.3–29.
210. Wentzel G., *Z. Physik*, 1927, Bd.43, S.1–8.
211. Wentzel G., *Z. Physik*, 1927, Bd.43, S.779–787.
212. DuMond J.W.M., *Rev. Mod. Phys.*, 1933, v.5, p.1–33.
213. Waller I., *Phil. Mag.*, 1927, v.4, p.1228–1237.
214. Weyrich W., Pattison P., Williams B.G., *Chem. Phys.*, 1979, v.41, p.271–284.
215. Kane P.P., Physics Reports, 1992, v.218, p.67–139.
216. Deb A., Chatterjee A.K. *Phys. Rev.*, 1996, B53, p.1393–1399.
217. Cooper M.J., *Radiat. Phys. Chem.*, 1997, v.50, p.63–76.
218. Coefficients of extinction of x-radiation (compiled by A.L. Zapysov, et al.,) Burevestnik, Leningrad, 1975.
219. Maximon L.C., Gimm H.A., *Phys. Rev.*, 1981, A23, p.172–185.
220. Hubbell J.H., Gimm H.A. and Overbo I., *J. Phys. Chem. Ref. Data*, 1980, v.9, p.1023–1147.
221. Belykh V.V., Smagunova A.N., Kozlov V.A., *Zh. Analit. Khimii*, 1994, v.49, No.10, p.1092–1096.
222. Guinier A., X-ray diffraction of crystals. Theory and practice, Fizmatlit, Moscow, 1961.
223. Friedrich W., Knipping P., Laue M., Interferenzerscheinung bei Rontgenstrahlen, Sitzungab. Bayer. Akad. Wiss. (Math.-Phys. Kl.), 1912, S. 303–322, 363–373.
224. Seemann H., *Phys. Zeitschr.*, 1919, Bd. 20, No.8, S. 169–175.
225. Debye P., Scherrer P., *Phys. Zeitschr.*, 1917, Jg.18, S. 291–301.
226. BraggW.H., Bragg W.L., *Proc. Roy. Soc. London* A, 1913, v.88, p.428–438.
227. Wulf G., *Phys. Zeitschr.*, 1913, Jg.14, S.217–220.
228. Landsberg G.S., Mandelstam L.I., *Die Naturwissenschaften,* 1928, Bd. 16, No.28, S.557–558.
229. Landsberg G.S., Mandelstam L.I., *Zh. Fiz. Khim. Obshch.*, 1928, v.60, p.335–338.
230. Raman C.V., Krishnan R.S., *Nature*, 1928, v.121, p.501–502.
231. Sparks C.J. jr., Anomalous scattering proc. intensity, Congress conf. organized by the Commission on Crystallographic Apparatus to the International Union of Crystallography held 22–26 April, 1974, p.175–191.
232. Cullie J., Sparks C.J.,jr., *Phys. Rev. Lett.*, 1974, v.33, p.262–265.
233. Eisenberger P., Platzman P.M., Winick H., *Phys.Rev.*Lett.,1976, v.36, p.623–626.
234. Kodre A.F., Shafroth S.M., *Phys. Rev.* A, 1979, v.19, No.2, p.675–678.
235. Pshenichnyi G.A., Raman scattering of X-radiation and elemental-phase analysis of matter, In: Equipment and methods of X-ray analysis, Mashinostroenie, Leningrad, 1988, No.37, p.24–55.

236. Briand J.P., Girard D., Chevalier P., et al., *Phys. Rev. Lett.*, 1981, v.46, No.25, p.1625–1629.
237. Tulkki J., *Phys. Rev.*, 1982, A27, No.6, p.3375–3378.
238. Pshenichnyi G.A. et al., Vestnik LGU, Ser. Geol. Geogr., 1982, No.24, p.33–43.
239. Russo D., Friedman G., Williams P., Raman scattering, In: Raman scattering of the light in gases and liquids, Mir, Moscow, 1982, p.247–309.
240. Landsberg G.S., Optics, 5th edition, Moscow, 1976.
241. Born M., Wolf E. Principles of optics. Electromagnetic theory of propagation, interference and diffraction of light, Pergamon Press, London, 6th ed., 1980.
242. Kallman H., Mark H., *Naturwiss.*, 1926, Bd.14, No.27, S.648–650.
243. Kallman H., Mark H., *Ann. d. Physik*, F.4, 1927, Bd.82, No.5, S.585–604.
244. Hohl H., *Ann. d. Physik*, F.5, 1933, Bd.18, No.6, S.625–655.
245. Izraileva L.K., Borovskii I.B., *Izv. AN SSSR, Ser. Fiz.*, 1972, v.36, No.2, p.438–450.
246. Klockenkamper R., Total-reflection X-ray fluorescence analysis (ed. Winefordner J.D.), John Willey & Sons Inc., New York, 1997.
247. Eikhenval'd A.A., Theoretical physics, vol. 6, GNTI, Moscow and Leningrad, 1931.
248. Sinaiskii V.M., In: Equipment and methods for X-ray analysis, Mashinostroenie, Leningrad, 1974, No..14, p.184–189.
249. Losev N.F., Krasnolutskii V.P., Losev V.N., *Zavod. Lab.*, 1993, v.59, No.6, p.20–29.
250. Kiessig H., *Ann. d. Phys.*, F.5, 1931, Bd.10, No.7, S.769–788.
251. De Boer D.K.G., van den Hoogenhof W.W., Adv. X-Ray Anal., New York, 1991, v.34, p.35–40.
252. De Boer D.K.G., Leenaers A.J.G., van den Hoogenhof W.W., *X-Ray Spectrometry*, 1995, v.24, p.91–102.
253. Abeles F., *Annalen de Physique*, 1950, v.5, p.596–640.
254. Krol A., Sher C.J., Kao Y.H., *Phys. Rev.*,1988, B38, p.85798592.
255. Bedzyk M.J., Bommarito G.M., Schildkraut J.S., *Phys. Rev. Lett.*,1989, v.62, p.1376–1379.
256. Zheludeva S.I., Kovalchuk M.V., Novikova N.N. et al., *Mater. Sci. Eng.*, 1995, C, C3(3–4), 211–14 (English), 1995.
257. Batterman B.W., *Phys. Rev.*, 1964, v.133, Issue 3A, p.A759–A764.
258. Batterman B.W. and Cole H., Rev. Mod. Phys.,1964, v.36,p.681–717.
259. Stoev K.N., Sakurai K., Spectrochim. Acta, 1999, B54, p.41–45.
260. Wobrauschek P., et al., Spectrochim. Acta, 1997, B52, p.901–906.
261. Egorov V.K., Kondratiev O.S., Zuev A.P., Egorov E.V., *Adv. X-Ray Anal.*, 2000, v.43, p.406–417.
262. Egorov V.K., Zuev A.P., Egorov E.V., *Zavod. Lab.*, 2001, v.67, No.3, p.3–11.
263. Vincze L., Janssens K., Adams F., Rindby A., Engstrom P., Rev. Sci. Instrum., 1998, v. 69, No.10, p.3494–3503.
264. Kumakhov M.A., *X-Ray Spectrometry*, 2000, v.29, No.5, p.343–348.
265. Dabagov S.B., Marcelli A., *Appl. Optics*, 1999, v.38, p.7494–7497.
266. Dabagov S.B., *X-Ray Spectrometry*, 2003, v.32, No.1, p.3–15.
267. Mazuritskii M.I., X-ray spectrum optics, Diapazon, Rostov-on-Don, 2005.
268. Blokhin M.A., Methods of X-ray spectrum investigations, Fizmatgiz, Moscow, 1959.
269. Riekel C., Bosecke P., Diat O., et al. Small-angle X-ray scattering at the European synchrotron radiation facility, J. Phys. IV (Paris), 1993, Col.C8, Suppl. JPI, No.12, IX International Conference on Small-Angle Scattering, p.403–410.
270. Kirkpatrick P., Baez V., *J. Opt. Soc. Am.*, 1948, v.38, p.766–774.

271. Howells M. R., Opt. Eng., 1995, v.34, p.410–417.
272. Chavanne J., et al., *Rev. Sci. Instrum.*, 1994, v.65, No.6, p.1959–1964.
273. Kuznetsov S.M., Snigireva I.I., Snigirev A.A., et al., *Appl. Phys. Lett.*, 1994, v.65, No.7, p.827–829.
274. Alexandrov Yu.M., et al., *Nucl. Instrum. Methods*, 1998, B134, p.174–180.
275. Kumakhov M.A., *Adv. X-Ray Anal.*, 1999, v.41, p.214–226.
276. Ibraimov N.S., et al., Investigation of the radiation integral intensity given by microfocus tube and X-ray lens, Proc. SPIE, 2000, v.4155, p.83–85.
277. Hosokawa Y., Ozawa S., Nakazawa H., Nakayama Y., *X-Ray Spectrometry*, 1997, v.26, p.380–387.
278. Auger P., Perrin F., *Journ. de Phys. et Rad.*, 1927, v.8, p.93–112.
279. Compton A.H., *Phil. Mag.*, 1929, v.8, p.961–977.
280. Auger P., *Ann. Phys.*, 1926, v.6, p.183–205, 205–221, 222–238, 239–253.
281. Hoogewijs R., Fiermens L., Vennik J., *Surf. Sci.*, 1977, v.69, p.273–294.
282. Brytov I.A., Ivanov V.Sh., Korablev V.V., 60 years since the discovery of the Auger effect, In: Equipment and methods for X-ray analysis, Mashinostroenie, Leningrad, 1988, No.37, p.3–24.
283. Carlson T., Photoelectronic and Auger spectroscopy [Russian translation], Mashinostroenie, Leningrad, 1981
284. Wentzel G., *Zeits. F. Phys.*, 1927, v.43, p.524–531.
285. Listengarten M.A., *Izv. AN SSSR, Ser. Fiz.*, 1961, v.25, No.7, p.792–798.
286. Listengarten M.A., *Izv. AN SSSR, Ser. Fiz.*, 1960, v.24, No.9, p.1041–1075.
287. Pincherle L., *Physica*, 1935, v.2, p.597–605.
288. Parilis E.S., The Auger effect, FAN UzbSSR, Tashkent, 1969.
289. Compton A.H. and Hubbard H., *Phys. Rev.*, 1924, v.23, p.439–449.
290. Pavlinsky G.V., Theoretical fundamentals of computational experiments in X-ray fluorescence analysis, In: Equipment and methods for X-ray analysis, St. Petersburg, 1992, No.41, p.83–104.
291. Sherman J., *Spectrochem. Acta*, 1955, 7(5), 283–295.
292. Makov V.M., Losev N.F., Pavlinsky G.V., *Zavod. Lab.*, 1968, v.34 , No.12, p.1459–1460.
293. Losev N.F., Glotova A.N., Afonin V.P., Zavod Lab., 1963, v.29, No.4, p.421–426.
294. Berry P., Furuta T., Rhodes J., *Adv. X-Ray Anal.*, 1969, v.12, p.612–632.
295. Mitra G.B., Wilson A.J.G., Brit. *J. Appl. Phys.*, 1960, v.11, No.1, p.43–45.
296. Khislavskii A.G., Plotnikov R.I., Kokhov E.D., *Zavod. Lab.*, 1977, v.43, No.9, p.1079–1082.
297. Shashilov V.P., Pavlinsky G.V., *Zavod. Lab.*, 1977, v.43, No.9, p.1079–1082.
298. Shashilov V.P., Vladimirova L.I., Pavlinsky G.V., *Zavod. Lab.*, 1979, No.11, p.1003–1008.
299. Hunter C.B., Rhodes S.R., *X-Ray Spectrometry*, 1972, v.1, No.3, p.107–111.
300. Duimakaev Sh.., Shpolyanskii A.Ya., Zhuravlev Yu.A., *Zavod. Lab.*, 1988, v.54, No.12, p.24–34.
301. Finkel'stein A.L., Gunicheva T.N, Kalughin A.G., Afonin V.P., *X-Ray Spectrometry*, 1992, v.21, No.6, p.287–292.
302. Finkel'stein A.L., Afonin V.P. Simplified algorithm of calculating the intensity of X-ray fluorescence of powder samples by the Monte Carlo method, In: Geochemical search methods. Analysis methods, Irkutsk, 1977, p.168–171.
303. Helsen J., Vrebos B., *Spectrochim. Acta,* 1984, v. 39B, No.6, p.751–759.
304. Druz V.V., Blokhin M.A., Aleinikov J.C., Die Methode der Eichung bei Rontgenfluoreszenzanalyse mehrkomponentigen System, Vortrage der Tagung 'Rontgenspectrometrie und ihre Bedeutung als Analysenmethoden' in Dresden.

Deutsche Gesellschaft fur Geologische Wissenschaft: Berlin, 1968, p.116–131.
305. Pavlinsky G.V., Kitov B.I., Tyumentsev V.N., Role of high-energy X-ray emission in excitation of L-fluorescence, In: Equipment and methods for X-ray analysis. Mashinostroenie, Leningrad, 1987, No. 36, p.49–53.
306. Bambinek W., Craseman B., Fink R.W., et al., *Rev. Mod. Phys.*, 1972, v.44, No.4, p.716–813.
307. Ebel H., Landler F., Dirschmid H., *Zeits. fur Naturforschung,* 1971, B.25a , H.5, s.927–928.
308. Pavlinsky G.V., Gulyaev V.T., Evaluation of the contribution of Auger photoelectrons to excitation of fluorescent X-radiation, In: Investigations in solid-state physics, Irkutsk, 1974, No.2, p.230–233.
309. Afonin V.P. Piskunova L.F., *Zavod. Lab.*, 1978, v.44, No.9, p.1083–1086.
310. Pavlinsky G.V., Vladimirova L.I., Role of photo- and Auger electrons in excitation of L-fluorescence, In: Equipment and methods for X-ray analysis. Mashinostroenie, Leningrad, 1988, No. 37, p.65–69.
311. Pavlinsky G.V., Formation of X-ray fluorescence of elements with low atomic number, Pakistan, Islamabad, The Nucleus, 1995, vol.32, Nos.3–4, p.119–124.
312. Philibert G., Some problems of quantitative X-ray spectrum local analysis, In: Physical fundamentals of X-ray spectrum local analysis, Nauka, Moscow, 1973, p.139–157.
313. Zalesskii V.V., *Optika i Spektroskopiya*, 1964, v.17, No.4, p.576–582.
314. Pollai I., Mantler M., Ebel H., *Spectrochim. Acta*, 1971, 26B, p.747–759.
315. Pavlinsky G.V, Pan'kov L.V., *Defektoskopiya*, 1993, No.12, p.72–82.
316. Pollai G., Mantler M., Ebel H., *Spectrochem. Acta*, 1971, v.26B, S.733–746.
317. Ebel H., Hillbrand E., *Spectrochem. Acta*, 1972, v.27B, No10, p.463–469.
318. Velichko Yu.I., Makhotko V.F., Revenko A.G., *Zavod. Lab.*, 1976, v.42, No.11, p.1338–1347.
319. Finkel'shtein A.L., Afonin V.P., Gunicheva T.N., *Zavod. Lab*, 1985, v.51, No.12, p.20–22.
320. Fernandez J.E., *X-Ray Spectrometry*, 1992, v.21, p.57–68.
321. Kharchenko A.M., Afonin V.P., Evaluation of the effect of additional excitation in X-ray spectrum fluorescent analysis by L-spectra, In: Equipment and methods for X-ray analysis. Mashinostroenie, Leningrad, 1977, No. 19, p.50–56.
322. Blokhin M.A., Methods of X-ray spectrum analysis, Thesis, Rostov State University, Rostov-on-Don, 1952.
323. Sherman J., The correction between fluorescent X-ray intensity and chemical composition, ASTM. Spec. Techn. Publ., 1954, No. 157, p.27–33.
324. Pavlinsky G.V., Losev N.F., *Zh. Tekh. Fiz.*, 1969, v.39. No. 9, p.1664–1675.
325. Pavlinsky G.V., Imeshenkova N.N., Evaluation of the second and third order effects and taking these effects into account in calculation of the intensity of X-ray fluorescence, In: Methods of X-ray spectrum analysis, Nauka, Novosibirsk, 1986, p.13–17.
326. Kuznetsova A.I., Lavrent'ev Yu.G., *Zavod. Lab.*, 1976, v.42, No.6, p.674–680.
327. Losev N.F., Pavlinsky G.V., *Zh. Tekh. Fiz.*, 1968, v.38, No.10, p.1803–1809.
328. Pavlinsky G.V., Kitov B.I., Losef N.F., *Zavod. Lab.*, 1974, v.40, No.9, p.1083–1087.
329. Pavlinsky G.V., *Zh. Anal. Khimii,* 1985, v.40, No.8, p.1407–1415.
330. Pavlinsky G.V., Wladimirova L.I., *X-Ray Spectrometry*, 1999, v.28, p.183–193.
331. Pavlinsky G.V., Losev N.F., *Zavod Lab.*, 1963, v.29, No.9, p.1067–1070.
332. Pavlinsky G.V., Losev N.F., Makov V.M., *Zavod. Lab.*, 1965, v. 31, No.9, p.1077–1081.

333. Tertian R., *Spectrochim. Acta*, 1971, v.26 B, No. 2, p.71–94.
334. Pavlinsky G.V., Kitov B.I., *Zavod. Lab.,* 1980, v.46, No.6, p.502–505.
335. Pavlinsky G.V., Kitov B.I., *Zavod. Lab*., 1982, v.48, No.4, p.21–24.
336. Pavlinsky G.V., Trubacheva T.A., Using the monochromatic model of characteristic L-radiation in calculations of the intensity of X-ray fluorescence, In: Equipment and methods of X-ray analysis, Mashinostroenie, Leningrad, 1983, No. 29, p.3–5.
337. Pavlinsky G.V., Increasing the accuracy of taking into account matrix effects in X-ray fluorescence analysis of multicomponent materials, Thesis, Moscow, 1989.
338. Pavlinsky G.V., Dukhanin A.Yu., *X-Ray Spectrometry*, 1994, v.23, p.221–228.
339. Pavlinsky G.V., Dukhanin AYu., *X-Ray Spectrometry*, 1995, v.24, p.293–297.
340. Dukhanin A.Yu., et al., *X-Ray Spectrometry*, 2006, v.35, No.1, p.34–39.

Index

R

S

T

U

W

X